THz for CBRN and Explosives Detection and Diagnosis

NATO Science for Peace and Security Series

This Series presents the results of scientific meetings supported under the NATO Programme: Science for Peace and Security (SPS).

The NATO SPS Programme supports meetings in the following Key Priority areas: (1) Defence Against Terrorism; (2) Countering other Threats to Security and (3) NATO, Partner and Mediterranean Dialogue Country Priorities. The types of meetings supported are generally "Advanced Study Institutes" and "Advanced Research Workshops". The NATO SPS Series collects together the results of these meetings. The meetings are co-organized by scientists from NATO countries and scientists from NATO's "Partner" or "Mediterranean Dialogue" countries. The observations and recommendations made at the meetings, as well as the contents of the volumes in the Series, reflect those of participants and contributors only; they should not necessarily be regarded as reflecting NATO views or policy.

Advanced Study Institutes (ASI) are high-level tutorial courses to convey the latest developments in a subject to an advanced-level audience.

Advanced Research Workshops (ARW) are expert meetings where an intense but informal exchange of views at the frontiers of a subject aims at identifying directions for future action.

Following a transformation of the programme in 2006, the Series has been re-named and re-organised. Recent volumes on topics not related to security, which result from meetings supported under the programme earlier, may be found in the NATO Science Series.

The Series is published by IOS Press, Amsterdam, and Springer, Dordrecht, in conjunction with the NATO Emerging Security Challenges Division.

Sub-Series

A.	Chemistry and Biology	Springer
B.	Physics and Biophysics	Springer
C.	Environmental Security	Springer
D.	Information and Communication Security	IOS Press
E.	Human and Societal Dynamics	IOS Press

http://www.nato.int/science
http://www.springer.com
http://www.iospress.nl

Series B: Physics and Biophysics

THz for CBRN and Explosives Detection and Diagnosis

edited by

Mauro F. Pereira
Materials & Engineering Research In
Sheffield Halam University
Sheffield, United Kingdom

and

Oleksiy Shulika
Campus Irapuato-Salamanca
Universidad de Guanajuato, Departamento de Ingeniería
Salamanca, Guanajuato, Mexico

Springer

Published in Cooperation with NATO Emerging Security Challenges Division

Proceedings of the NATO Advanced Research Workshop on THz for CBRN and Explosives Detection and Diagnosis
Çeşme, Turkey
3–5 November 2015

Library of Congress Control Number: 2017945743

ISBN 978-94-024-1095-2 (PB)
ISBN 978-94-024-1092-1 (HB)
ISBN 978-94-024-1093-8 (eBook)
DOI 10.1007/978-94-024-1093-8

Published by Springer,
P.O. Box 17, 3300 AA Dordrecht, The Netherlands.

www.springer.com

Printed on acid-free paper

Preface

Hazardous substances typically exhibit rotational and vibrational transitions in this region, hence giving access to spectroscopic analysis of a large variety of molecules which play a key role in security as well as various other areas, e.g. air pollution, climate research, industrial process control, agriculture, food industry, workplace safety and medical diagnostics can be monitored by sensing and identifying them via THz (0.3–10 THz) and mid-infrared (MIR-10–100 THz) absorption "fingerprints". Most plastics, textiles and paper are nearly transparent for THz radiation.

Therefore, CBRN agents, explosives and illegal drugs can be detected by their characteristic absorption spectra at THz frequencies with high selectivity and resolution in application fields as industrial quality inspection control, customs inspection and security screening. Moreover, THz and MIR radiation has no endangering effects on human beings and enables higher contrast for "soft matter" than X-rays. In comparison to standard optical technologies for wavelengths up to about 2 μm, sources and detectors for THz and MIR have not yet reached their level of maturity, and there is still a large gap for features like wavelength tunability, spectral purity, high power and room temperature operation, which all are necessary for commercial applications; thus research and development are strongly needed.

Plastic or ceramics are detected by X-rays very poorly especially against a background of human body. Unlike X-rays, TERA-MIR is not a dangerous radiation, and in some cases THz sensors can reveal not only the shape of a hidden object but also its chemical composition. This unique combination of traits makes THz and mid-infrared perfect for effective applications like explosive detection and security applications. Besides, THz radiation has a high resolution in a 3D space in case of ultrashort pulses. The identification of hidden hazardous substances under paper and cloth is possible with devices operating in the THz and mid-infrared ranges. However, there are still bottlenecks to be overcome for widespread applications of this technology.

In order to cover the most relevant topics in the field, the NATO ARW THz for CBRN and Explosives' Detection and Diagnostics, which took place in Çeşme, Turkey, on 3–5 November 2015, started with a session focused on security and detection of CBRN and explosives, including dual-use diagnostics. For example, some techniques were discussed for the detection of different objects and substances. For the first time in our series of NATO ARWs dedicated to THz and its applications, we had a participant directly involved in "Situational Awareness Frameworks for National Security and Protection of National Critical Security – How Can Early Detection of Threats by THz Devices Help Our Efforts?" This was crucial to help the participants have new ideas on how to connect the relevant science and technology to security agencies. Next, we had detection talks: Innovative Quartz Enhanced Photoacoustic Sensors for CBRNs Detection Operating in THz and Mid-Infrared Spectral Ranges; Terahertz Plasma FETs, from Basic Physics to First Fast Terahertz Scanners for Detection of Explosives and CBRN; THz Transceivers and Their Application to Detection of CBRN and Explosives; Integrating THz Sensors/Structures through Electro-wetting in Dielectrics (EWOD) for Security Applications; THz Detection and Identification of Dangerous Objects and Substances; and Compressive Sensing for Detection of Concealed Objects. This was followed by a poster session with presentations covering different types of sources and detectors of THz as well as their applications to sensing.

The main THz and MIR source is the quantum cascade laser (QCL). A strong emphasis was given to invited talks from leading scientists from leading teams dedicated to turning this advanced technology into affordable commercial devices throughout the THz spectral range and exploiting their enormous potential for security applications in the second day: World's First mW Power, Room Temperature, Continuous Wave 3.5 THz Diode Laser, based on QCL Difference Frequency Generation; Recent Progress on Terahertz Quantum Cascade Lasers; Advances in QCLs Optimized for Chemical Sensing; Phase-Locked Spectroscopy of THz QCLs; Development and Applications of THz Frequency QCLs; Recent developments in QEPAS: a spectroscopic technique for trace gas sensing; and Microscopic Theory for Interaction of THz Radiation with Semiconductor Materials – from Bulk to Nanostructures.

Regardless of the importance of QCLs, we discussed alternative THz sources in the session: Beyond Quantum Cascade Lasers: Competing Materials and Technologies for THz and MIR Radiation Detection and Applications. The following topics were discussed: THz Field Induced Second Harmonic Coherent Detection Scheme Based on a Biased Nonlinear Micro-slit, Sensitivity and Selectivity in THz Spectroscopy of Bioparticles and Biomolecules, THz Emission from Bulk and Nanowire Semiconductor Surfaces, THz Flux-Flow Oscillators for Detection of Explosive and CBRN Agents and Sensitive Terahertz-wave Detection Using Optical Nonlinear Wavelength-conversion for Stand-off Sensing. The session continued in the last day with: Terahertz Wave Emission and Detection from Single Crystals of High Temperature Superconducting $Bi_2Sr_2CaCu_2O_{8+\delta}$, THz Wave Sensing @ ETRO-VUB: Beyond Conventional Limits, CBRN Defence Using THz Pulse Trains from Semiconductor Disk Lasers, Compact photomixing solutions for generation/detection of mm/THz-waves: monolithic photonic circuits and self-coherent detection, Method for Generation Multiple High-Intensity THz Pulses within 1–100 ps Time Range, Design Aspects of Antennas for THz Bolometers, Far Infrared System for Detection of Explosives and CBRN Agents for Post Office and THz Spectroscopy for Detection of Explosives and Illicit Substances.

We closed the event with a half-day devoted to formulating workshop conclusions and recommendations, starting with a roundtable in THz and MIR applications in security area problems/possible science for peace proposals. This discussion took place hand in hand with the forum "From Academia to Industry" in which research-intensive industrial partners proposed pathways for turning fundamental science presented in the workshop into concrete development projects for commercialization.

This book gives more details about some of the topics presented in the workshop, including sensing techniques and systems, properties of the materials required for new devices and THz sources.

In summary, the presentations delivered a global review of the state of the art on basic science, technology and applications of THz for CBRN and explosives. Joint research and proposal applications for cooperation projects have been submitted based on discussions that started during this meeting.

Sheffield, UK
Salamanca, Mexico
November, 2016

Mauro Fernandes Pereira
Oleksiy Shulika

Acknowledgements

We start our acknowledgement list with our thanks to NATO and the Science for Peace project for the very generous financial support and continuous support with all necessary details which made the realization of the NATO Advanced Research Workshop on Detection of Explosives and CBRN (Using Terahertz), which took place in Çeşme, Izmir, Turkey, on 3–6 November 2015, and this book possible. The co-chairs Mauro Pereira and Lilian Sirbu are grateful to NATO, which provided a fantastic opportunity for all of us to meet in Çeşme, Izmir. Joint proposals and scientific collaborations are already evolving thanks to this opportunity, and we hope that they will have an impact in the development of research in the fascinating TERA-MIR range.

Our big thanks to Kim Turner and the finance team at Sheffield Hallam University who helped us go through the many financial details needed to organize this meeting and for taking care of the travel and accommodation for colleagues arriving from around the world very efficiently and to Emilia Pruszyńska-Karbownik with her support for our website.

The other committee members played a major role in helping us select the speakers and reach a final programme conclusion, so here is a statement of our appreciation for the support given by Lilian Sirbu, Lutfi Oziuzer and his team, Anna Wójcik-Jedlińska, Czeslaw Skierbiszewski, Shumin Wang, Antoine Muller, Andreas Wacker, Ivan Buchvarov, Johan Stiens and Florin Mihailescu.

In preparing this book, we have relied on the contribution of the authors. Without their expert insight, motivation and commitment, the publication of this volume would not have been possible. We, thus, extend our appreciation to all the authors. We also convey our thanks to Springer for the opportunity of publishing this volume.

Contents

Part I
Sensing Techniques and Systems

Chapter 1
Toward Standoff Sensing of CBRN with THz Waves

Kang Liu and Xi-Cheng Zhang

Abstract With the soaring demands for remote spectroscopy in homeland security and environmental monitoring, terahertz (THz) wave sensing has drawn a significant amount of attention because of its capability to acquire chemical spectral signatures non-invasively. THz-Radiation-Enhanced-Emission-of-Fluorescence (THz-REEF) makes broadband stand-off remote spectroscopy possible, due to its omni-directional emission pattern of the fluorescence and minimal ambient water vapor absorption of THz wave near the target. Two years ago, we have demonstrated coherent THz wave detection at a distance of 30 m. The development of intense broadband table-top THz source is imperative to all the extreme THz science research, including THz remote sensing. Recently, we have also demonstrated an enhanced THz radiation using dual-color ring-Airy beam induced air-plasma compared to a Gaussian beam induced plasma. It is believed that the great tunability of ring-Airy beam, its wavepacket stability at nonlinear focus regime, and the strong resilience to turbulent medium of the Airy beam family, make the plasma induced by this novel wave a promising broadband THz source for THz remote sensing and spectroscopy.

Keywords Terahertz • Terahertz standoff sensing • Terahertz generation • Plasmas • Ring-airy beams

1.1 Introduction: THz Air-Photonics

Since the THz generation process in air plasma is intimately tied with gas ionization, it requires that intensities greater than 10^{12} W/cm^2 be created near the target location. While this can be achieved through geometric focusing with lenses or curved mirrors, it can also be accomplished with a technique called filamentation [1]. Filamentation is the process by which a high-intensity beam self-focuses through nonlinear processes and collapses, forming a plasma channel with mm to

K. Liu • X.-C. Zhang (✉)
The Institute of Optics, University of Rochester, Rochester, NY 14627, USA
e-mail: xi-cheng.zhang@rochester.edu

M.F. Pereira, O. Shulika (eds.), *THz for CBRN and Explosives Detection and Diagnosis*, NATO Science for Peace and Security Series B: Physics and Biophysics,
DOI 10.1007/978-94-024-1093-8_1

cm lengths. This plasma channel stabilizes the beam at very small diameters (30–100 μm) and maintains high intensities over ranges much longer than the Rayleigh length of a traditional, geometrically focused beam.

Similar to the widely used generation and detection of THz waves in electro-optic crystals by second-order optical nonlinearity, the THz waves can be detected by the third-order optical nonlinearity in air. This THz wave air-plasma photonics system [2, 3], which differentiates itself from most THz time-domain spectroscopy systems [4–8], uses ambient air or selected gases for both generation and detection of broadband THz waves.

Figure 1.1a illustrates this all-optical process for the generation and detection of THz waves using an air emitter and sensor. A Ti:sapphire regenerative amplifier (~3 mJ, 30 fs and 1 kHz repetition rate with a central wavelength at 800 nm) is used as a laser source. The laser beam is focused through a 100-μm thick type-I beta barium borate (BBO) crystal to generate its second harmonic beam. The mixed fundamental (ω) and its second harmonic beams (2ω) generate a THz wave at the ionizing plasma spot. The THz beam is recombined with the fundamental at a second focus, creating another plasma spot, which serves as the detection region. An electric bias field is applied to the detection region, supplying a second harmonic local oscillator for coherent detection via a photomultiplier detector [9]. The plasma emits a very intense, highly directional, ultra-broadband THz field.

Figure 1.2a,b plot a typical temporal THz waveform and its spectrum from laser-induced air plasma, using THz air-biased coherent detection (THz-ABCD), in comparison with the signal detected by ZnTe [10]. Air-plasma generated THz waveform measured by THz-ABCD can have a peak THz electric field exceeding 1 MV/cm with a dynamic range over 3000. Its bandwidth covers the THz gap (0.3–10 THz). Now THz-ABCD becomes a standard THz wave detection technique for researchers involved in projects investigating ultra-broadband THz pulses and their applications.

1.2 Remote Sensing with THz Air-Photonics

1.2.1 Remote THz Wave Generation in the Ambient Air

By focusing intense two-color beam (800 nm and 400 nm) in ambient air, we are able to generate THz wave remotely by the induced plasma. To achieve remote generation, a fine control of the phase of the two colors is required, due to chromatic dispersion in air. Therefore, a device used to compensate this walk off is the key to standoff THz wave generation, which is called a "phase-compensator". Thanks to this technique, THz wave generation at distances up to 30 m from the laser source has been demonstrated [11] (Fig. 1.3).

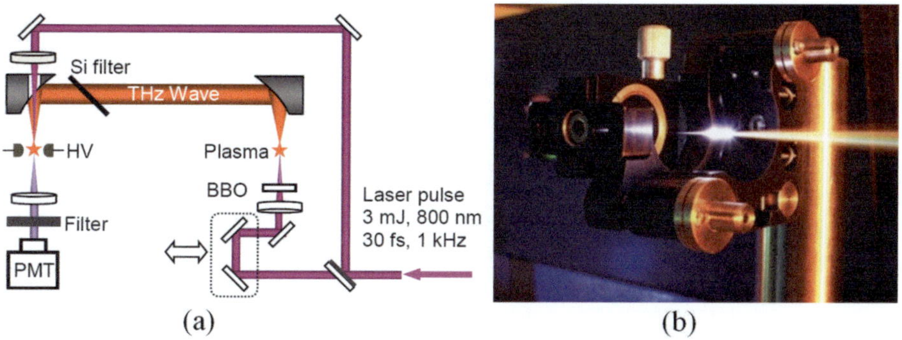

Fig. 1.1 (**a**) Schematic diagram of the THz-ABCD design concept. (**b**) Laser-induced air plasma (*center bright line*)

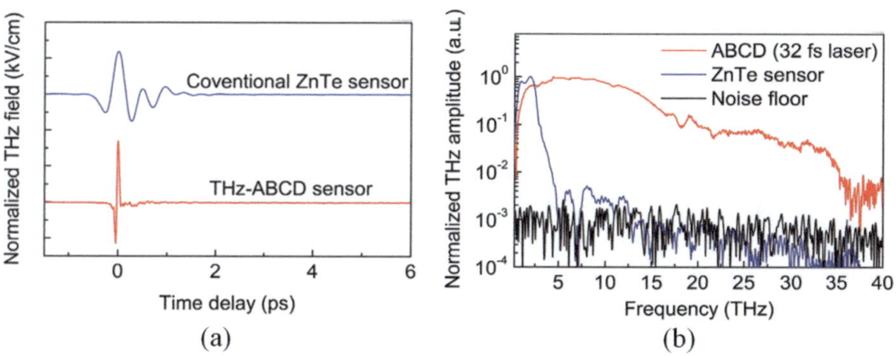

Fig. 1.2 (**a**) Time-resolved THz signals generated and detected using dry nitrogen gas as compared to conventional EO crystal detection in ZnTe. The probe beam for air detection has energy of 85 μJ and pulse duration of 32 fs. (**b**) Corresponding spectra after Fourier transform

1.2.2 Remote THz Wave Detection with THz-REEF

THz wave signals can be detected indirectly by observing THz Radiation Enhanced Emission of Fluorescence (THz-REEF). In fact, when a THz pulse is spatially and temporally overlapped with a laser induced plasma, the spectral information it contains is encoded in its fluorescence emission (wavelength of 300–400 nm), which experiences negligible atmospheric absorption [12].

Using a two-color beam to form the plasma sensor, the information about the time-dependent terahertz field can be directly retrieved by taking the differential between $\Delta I_{FL}(\Delta\varphi_{\omega,2\omega} = \pi/2)$ and $\Delta I_{FL}(\Delta\varphi_{\omega,2\omega} = -\pi/2)$, where $\Delta\varphi_{\omega,2\omega}$ is the relative phase of the two optical beams [13], as shown in Fig. 1.4.

Recently, we demonstrated coherent terahertz wave detection at a distance of 30 m. Figure 1.5a shows a schematic figure of the experimental set-up. A pair of wedge is used to control the relative phase $\Delta\varphi_{\omega,2\omega}$ between ω and 2ω. The THz

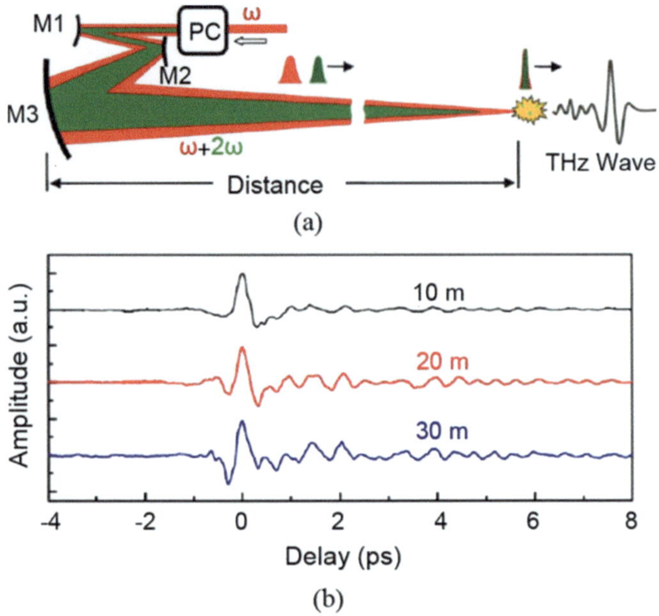

Fig. 1.3 (**a**) Schematic demonstration of remote THz wave generation from laser-induced plasma in ambient air; M1, convex spherical mirror used to expand the optical beams; and M2, concave spherical mirror used to remotely focus optical beams. (**b**) Three THz waveforms generated in ambient air at distances of 10, 20 and 30 m, respectively, using 1-mm-thick ZnTe crystal as the sensor

Fig. 1.4 Typical THz-REEF signal. *Blue* and *green*: two THz-REEF traces with different relative phase $\Delta\varphi_{\omega,2\omega}$. *Red*: the difference between *blue* and *green curves*, that gives the measured THz waveform

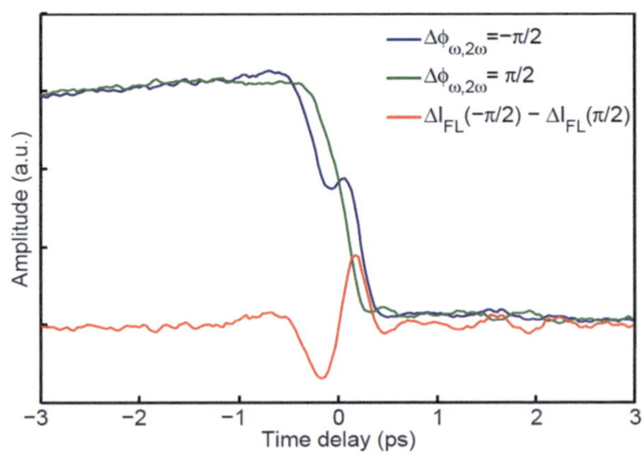

pulse and the optical pulses were focused collinearly onto one spot by two off-axis parabolic mirrors. The plasma fluorescence radiation which contains the THz signal is collected by a big spherical mirror and refocused to a PMT detector. The distance between the plasma and the spherical mirror is the detection distance of the remote sensing. Shown in Fig. 1.5b, a signal-to-noise dynamic range of ~4 was achieved

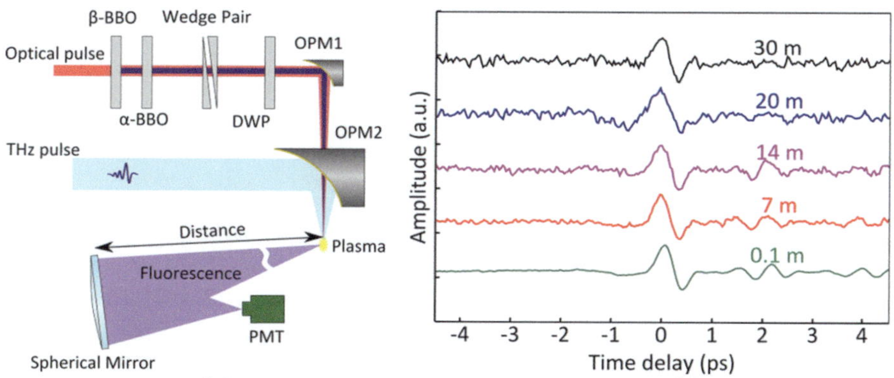

Fig. 1.5 (**a**) Schematic figure of THz-REEF remote sensing experimental setup; *DWP*: dual wave plate; *OPM*: off-axis parabolic mirror, *PMT*: photomultiplier tube detector; (**b**) THz waveforms measured by THz-REEF technique at different detection distances, from 0.1 to 30 m

after 64 averages for the 30 m signal. This indicates all air-plasma stand-off THz wave remote sensing at 30 m is feasible in the lab.

1.2.3 Broadband THz Wave Remote Sensing Spectroscopy

With the development of remote THz wave generation technique and THz-REEF remote sensing technique, it is thus possible to envision a standoff detection scheme where a laser field induced air-plasma acts as a remote THz wave emitter, while another air-plasma induced by a second laser field acts as a remote THz wave sensor. Stand-off detection of dangerous chemicals, like explosive compounds is no longer hard to achieve [14] (Fig. 1.6).

1.3 THz Wave Generation Using Ring-Airy Beams

In the THz community, the development of table-top intense broadband THz source has always been a topic with great importance. One of the most efficient approaches is using two-color femto-second laser plasma [2, 15]. Ever since this method was demonstrated, it has caused a substantial amount of discussion and people have tried to play with different parameters of plasma to improve the generation efficiency.

It is believed that the great tunability of ring-Airy beam [16] and its wavepacket stability at nonlinear focus regime [17] make the plasma induced by this novel wave a promising broadband THz source with interesting potentials, such as in THz remote sensing spectroscopy [11].

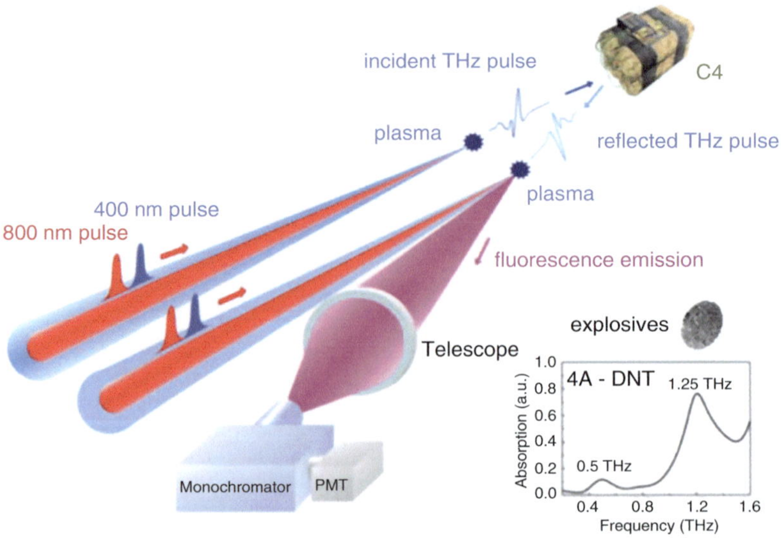

Fig. 1.6 Envisioned scheme for THz Stand-off generation and detection. Two dual-color pulses are focused close to the target under investigation creating a plasma emitter and a plasma sensor. Inset shows an absorbance spectrum of 4A-DNT retrieved through THz-REEF

In our experiment, the ring-Airy beam was generated by sending a 35 fs, 50 Hz, 800 nm Ti-Sapphire laser beam to a Spatial Light Modulator (SLM) with a phase mask exerted on it. The unwanted zero-order in the center was blocked to give the ring-Airy distribution, see Fig. 1.7a. We measured the THz waveform using free space electro-optical sampling with a 3-mm-thick ZnTe crystal. A pair of off-axis parabolic mirrors were used to collect and refocus the THz radiation from plasma. A CCD camera took the fluorescence images of the plasmas.

In Fig. 1.7c, d, THz waveforms and the corresponding spectrums emitted from the two plasmas in Fig. 1.7b are plotted together. During the experiment, the lens location as well as the β-BBO angle and location were all optimized for each measurement to avoid the signal discrepancy caused by focus location changes and the second harmonic generation efficiency difference. When we did the integration over time on the square of the waveforms in Fig. 1.7c in order to retrieve the THz pulse energy, we found that the THz pulse energy generated by the ring-Airy beams contains about 4.7 times the energy of the one by Gaussian, which is possibly due to the plasma length difference between the two beams [18]. At the same time, the ring-Airy THz emission has a center frequency slightly lower than the Gaussian one, which matches the fact that the Gaussian plasma has higher plasma density [19].

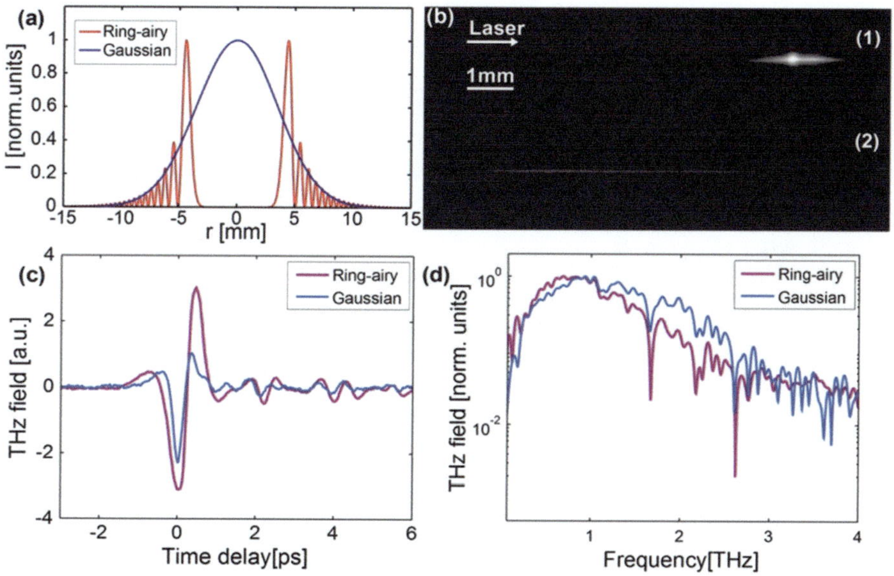

Fig. 1.7 (**a**) Schematic comparison between the radial intensity distribution of the ring-Airy and the Gaussian beam used in the experiment. (**b**) Fluorescence emission images of (1) Gaussian beam and (2) ring-Airy beam two color air-plasmas in *grey* color. The intensity is shown in a logarithmic scale to make both plasmas visible. (**c**) THz waveform generated by the ring-Airy plasma (*pink*) in comparison with the one generated by Gaussian plasma (*blue*). (**d**) Normalized THz spectrums emitted by the ring-Airy plasma (*pink*) in comparison with the one generated by Gaussian plasma (*blue*)

1.4 Conclusion

In conclusion, we have demonstrated both remote THz-REEF detection and remote THz wave generation at a distance of 30 m, which indicates all air-plasma stand-off THz wave remote sensing at 30 m is feasible in the lab. With current dynamic range, taking into account the reflection loss of the UV mirrors we used in order to create 30 m distance in the lab, and the fact that the signal drops quadratically with the distance, the potential maximum distance that we can do THz-REEF remote detection should be ~100 m. We have also shown the possibility of using ring-Airy beam to generate THz wave from laser-induced plasma in the ambient air, with a comparable efficiency as normal Gaussian beam. We measured that under the same experimental conditions, there is a small discrepancy between the THz pulse energy and spectrum generated by ring-Airy beam and Gaussian beam, possibly due to the length and electron density difference of the corresponding plasmas. We believe these results pave the way of extending ring-Airy beams as well as other exotic beams applications into THz frequency. It sets a very useful reference for the scientists who will explore using different novel wavepackets as THz sources in the future.

References

1. A. Couairon, A. Mysyrowicz, Femtosecond filamentation in transparent media. Physics Reports 441(2–4), 47–189 (2007). doi:10.1016/j.physrep.2006.12.005
2. X. Xie, J. Dai, X.-C. Zhang, Coherent Control of THz Wave Generation in Ambient Air. Phys. Rev. Lett. 96, 075005 (2006). doi:10.1103/PhysRevLett.96.075005
3. J. Dai, X. Xie, X. C. Zhang, Detection of Broadband Terahertz Waves with a Laser-Induced Plasma in Gases. Physical Review Letters 97(10), 103903 (2006). doi:10.1103/PhysRevLett.97.103903
4. D. J. Cook, R. M. Hochstrasser, Intense terahertz pulses by four-wave rectification in air. Optics Letters 25, 1210 (2000). doi:10.1364/OL.25.001210
5. S. Tzortzakis, G. Méchain, G. Patalano, Y. B. André, B. Prade, M. Franco, A. Mysyrowicz, J. M. Munier, M. Gheudin, G. Beaudin, P. Encrenaz, Coherent subterahertz radiation from femtosecond infrared filaments in air. Optics Letters 27(21), 1944–1946 (2002). doi:10.1364/OL.27.001944
6. M. Kress, T. L. ler, S. Eden, M. Thomson, H. G. Roskos, Terahertz-pulse generation by photoionization of air with laser pulses composed of both fundamental and second-harmonic waves. Optics Letters 29(10), 1120 (2004). doi:10.1364/OL.29.001120
7. T. Bartel, P. Gaal, K. Reimann, M. Woerner, T. Elsaesser, Generation of single-cycle THz transients with high electric-field amplitudes. Optics Letters 30(20), 2805–2807 (2005). doi:10.1364/OL.30.002805
8. A. Nahata, T. F. Heinz, Detection of freely propagating terahertz radiation by use of optical second-harmonic generation. Optics Letters 23(1), 67–69 (1998). doi:10.1364/OL.23.000067
9. N. Karpowicz, J. Dai, X. Lu, Y. Chen, M. Yamaguchi, H. Zhao, X.-C. Zhang, L. Zhang, C. Zhang, M. Price-Gallagher, C. Fletcher, O. Mamer, A. Lesimple, K. Johnson, Coherent heterodyne time-domain spectrometry covering the entire "terahertz gap". Applied Physics Letters 92(1), 011131 (2008). doi:10.1063/1.2828709
10. B. Clough, J. Dai, X.-C. Zhang, Laser air photonics: beyond the terahertz gap. Materials Today 15(1–2), 50–58 (2012). doi:http://dx.doi.org/10.1016/S1369-7021(12)70020-2
11. J. Dai, J. Liu, X.-C. Zhang, Terahertz Wave Air Photonics: Terahertz Wave Generation and Detection With Laser-Induced Gas Plasma. IEEE Journal of selected topics in Quantum Electronics 17, 183 (2011). doi:10.1109/JSTQE.2010.2047007
12. J. Liu, X. C. Zhang, Terahertz-Radiation-Enhanced Emission of Fluorescence from Gas Plasma. Physical Review Letters 103(23), 235002 (2009). doi:10.1103/PhysRevLett.103.235002
13. J. Liu, J. Dai, S. L. Chin, X. C. Zhang, Broadband terahertz wave remote sensing using coherent manipulation of fluorescence from asymmetrically ionized gases. Nature Photonics 4(9), 627–631 (2010). doi:10.1038/nphoton.2010.165
14. B. Clough, J. Liu, X. C. Zhang, "All air-plasma" terahertz spectroscopy. Optics Letters 36(13), 2399–2401 (2011). doi:10.1364/OL.36.002399
15. E. Greenfield, M. Segev, W. Walasik, O. Raz, Accelerating Light Beams along Arbitrary Convex Trajectories. Phys. Rev. Lett. 106, 213902 (2011). doi:10.1103/PhysRevLett.106.213902
16. I. Chremmos, N. K. Efremidis, D. N. Christodoulides, Pre-engineered abruptly autofocusing beams. Optics Letters 36, 1980 (2011). doi:10.1364/OL.36.001890
17. P. Panagiotopoulos, D. G. Papazoglou, A. Couairon, S. Tzortzakis, Sharply autofocused ring-Airy beams transforming into non-linear intense light bullets. Nature Communications 4, 2622 (2013). doi:10.1038/ncomms3622
18. Y. S. You, T. I. Oh, K. Y. Kim, Off-Axis Phase-Matched Terahertz Emission from Two-Color Laser-Induced Plasma Filaments. Phys. Rev. Lett. 109, 183902 (2012). doi:10.1103/PhysRevLett.109.183902
19. K. Y. Kim, A. J. Taylor, J. H. Glownia, G. Rodriguez, Coherent control of terahertz supercontinuum generation in ultrafast laser–gas interactions. Nature Photonics 2, 605 (2008). doi:10.1038/nphoton.2008.153

Chapter 2
THz Wave Sensing at ETRO-VUB: Beyond Conventional Limits

V. Matvejev, A. Pourkazemi, H. Alkorre, G. Shkerdin, and J. Stiens

Abstract Here, we demonstrate a novel Terahertz (THz) wave characterization technique and modulation devices developed at ETRO-IR (VUB) that facilitate the detection of Chemical Biological Radiation and Nuclear (CBRN) hazards. First, a novel transient radar technique is presented for a Non-Destructive Testing (NDT) of a multilayered dielectric object. A transient reflected electromagnetic wave is captured with femtosecond time resolution, so it can be able to reconstruct both the dielectric and thickness information from multitude of scattered waves from different dielectric interfaces in a multilayered object. A blind reconstruction algorithm does not require any priori knowledge of the object under test. Proof-of-principle is demonstrated experimentally at 10 GHz on a three-layered object, featuring subwavelength depth resolution. Second, amplitude and phase modulators are proposed based on hollow metal pipe waveguide loaded with single layer graphene on a dielectric substrate. It is demonstrated that the modes propagation constants can be substantially modulated by the graphene Fermi level modulation especially near the regions of strong mode modifications. Solutions of wave propagation equations are given and design parameters are optimized for implementing an efficient Terahertz modulator devices. Fast 3D object scanning and detection of CBRN hazards can be implemented in THz by means of transient radar with phased antenna array with graphene phase shifting modulators.

V. Matvejev (✉) • A. Pourkazemi • H. Alkorre
Department of Electronics and Informatics (ETRO-IR), Vrije Universiteit Brussel (VUB), Brussels, Belgium

G. Shkerdin
Kotel'nikov Institute of Radio Engineering and Electronics of the Russian Academy of Science, Vvedenskogo Square 1, 141120 Fryazino, Moscow, Russia

J. Stiens
Department of Electronics and Informatics (ETRO-IR), Vrije Universiteit Brussel (VUB), Brussels, Belgium

SSET, IMEC, Kapeldreef 75, 3001 Leuven, Belgium
e-mail: jstiens@etro.vub.ac.be

M.F. Pereira, O. Shulika (eds.), *THz for CBRN and Explosives Detection and Diagnosis*, NATO Science for Peace and Security Series B: Physics and Biophysics, DOI 10.1007/978-94-024-1093-8_2

Keywords Terahertz • Non-destructive testing • Transient measurement • Graphene loaded waveguide

2.1 Introduction

The research related to imaging and sensing at the GHz-THz laboratory of ETRO-VUB spans a broad range of activities: from theoretical electromagnetic concepts and material research up to devices, circuits and system design. Fast active multi-parameter MMW imaging systems for concealed weapon detection on the human body could perform the job in a few seconds, multi-parameter imaging enhancement techniques have been developed, functional antenna arrays have been implemented and characterized up to 220 GHz, meta-material characterization including sub-micron thickness sensing of thin coatings and materials, forward and inverse problem solving of multi-layer dielectric structures, the characterization of liquid mixtures at nano-liter level with a world record bio-sensor geometry (115 dB sensor signal change between water and ethanol). The group also developed skills in sub-wavelength near-field imaging with resolution of $\lambda/6000$. Recently a novel radar technique has been developed to obtain world record deep sub-wavelength resolution, $\lambda/100$. Compressive sensing techniques complement the radar work.

Due to the wide variety of industrial materials, there is a need for a complementary set of NDT techniques. Radiography is measuring changes in density [1]. Ultrasound techniques cannot deal with rubber or soft plastics, due to high absorption and negligible reflection of the material [2]. A viable alternative measurement technique for these industrial materials can be found in the microwave domain for extracting simultaneously the thicknesses and EM characteristics of each layer in a multi-layer stack. Several methods were proposed to calculate the dielectric properties of a multilayer structure and geometrical parameters. However, traditional algorithms developed for the EM spectrum, comprising the microwave and mm wave range, are time consuming, require a high frequency range when targeting high spatial resolution, or need a priori knowledge, to extract in an accurate way the properties of sub-mm thin layers. Various methods have been employed to determine the complex dielectric properties of materials from measured refection and transmission coefficients. The most common procedure to obtain the complex permittivity and permeability from S-parameters was proposed by Nicolson and Ross in 1970 [3], and modified by Weir in 1974 [4]. It is called Nicolson-Ross-Weir method. This method requires an ultra-wide frequency band and high frequency resolution to extract permittivity or permeability for a single layer. Finding characteristics of very thin single layer of material is big challenge for this algorithm [5, 6]. Synthetic Aperture imaging, which is based on Frequency-Modulated Continuous-Wave, needs ultra-width band frequency as well. This method needs prior knowledge for increasing accuracy [7]. In the millimeter wave spectrum, it yields a depth and lateral resolution of a few millimeters [8, 9]. Another method which is used for geometrical and distance is Time Domain Reflectometry (TDR). For small size and also thin layers millimeter or THz range will be used. This method uses pulses and works based on time-of-flight which is complex and needs expensive systems [10]. Hence finding novel methods/algorithm(s) which can cover all or part of these problems is useful. In the next paragraph, we present the novel method coined transient radar method (TRM) [11].

The remarkable electrical and optical properties of graphene in the mm and THz wave range make this material suitable for the implementation of reconfigurable circuits by simply applying an electrostatic biasing field or for the implementation of highly efficient modulators for complex modulation formats [12]. However, due to the nature of the physical mechanism, phase and amplitude variation never act independently. Latter research on graphene has shown the potential to deal with all modulator's challenges [13]. Coupled graphene-metal plasmons and waveguide modes for planar metal waveguide containing air/graphene monolayer/dielectric buffer layer/metal substrate) behavior were described at the sub-THz range [14, 15]. Furthermore, the dispersion equation of these waveguide modes of rectangular metal waveguide containing buffer/graphene/air gap layered structure has been derived in the approximation of perfectly conducting waveguide metal walls [16]. For the design of real waveguide modulator devices, it is essential to consider the finite length of dissimilar waveguide sections. Due to the discontinuities, the original modes in each section are no longer satisfactory solutions. For the correct prediction of the propagation of EM waves inside the full structures, it is essential to represent the novel solutions in each section as a (infinite) linear combination of modes, whether evanescent or propagating and to satisfy the boundary conditions at the discontinuities and exploiting mode orthogonality.

2.2 Transient Radar

In the Transient Radar Method (TRM) a multilayered structure (MLS) is periodically illuminated in free space with a monochromatic continuous electromagnetic wave and the time dependent transient reflected signal is captured by a narrow-band receiver. The TRM analyses the reflected transient signals caused by the rising or falling edge of a step function of the illumination wave. The concept of the TRM method is like the Time Domain Reflectometry as it deals with the analysis of propagation of a step function instead of a pulse. The reflected transient signal contains enough information to extract complex valued dielectric permittivity and thickness at the same time.

The TRM illumination configuration is shown in Fig. 2.1. The Proof-of-Concept is shown with this configuration at 10 GHz but can be easily scaled to higher frequencies. The illumination wave is periodically switched on and off by a fast "Single Pole Single Throw" switch (short rise time). The switch is placed between a MMW source and antenna. The reflected wave is received by the same antenna and coupled to a sampling oscilloscope, which captures the signal with 50 fs time

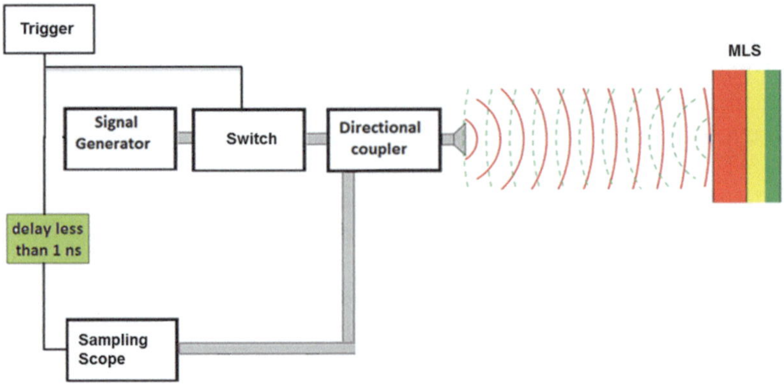

Fig. 2.1 TRM illumination configuration for NDT of a multilayered structure

resolution. The switch and sampling oscilloscope are synchronized with an indirect trigger and a short delay.

The TRM reconstruction algorithm extracts recursively layer by layer, starting from the closest layer to the illumination antenna, the complex dielectric permittivity and thickness from the transient reflection signal. The transient reflected signal represents a multitude of superposed wave fronts arriving at different instances of time that were scattered on different boundaries of the MLS, before the steady state is achieved. For one boundary one can easily calculate the permittivity from the impedance mismatch (or reflection coefficient) and distance to the boundary from the time of flight of the wave front. In case of MLS, one can calculate permittivity and thicnkess recursively for each subsequent wave front by considering the information for boundaries that are already solved. This enables to extract dielectric permittivity and thickness for each layer in MLS, although with progressively reduced accuracy for deeper layers due to the accumulation of errors.

The non-idealities and noise of the components needs to be compensated to get adequate TRM extraction parameters for MLS. First of all, calibration must be completed. By measuring without any object in front of the illuminator one can compensate for the cross-talk between the source and sampling oscilloscope (see Fig. 2.2a, d – green curve). By putting a metal reflector, it is possible to determine the switch rise time (see Fig. 2.2b, d – red curve) and the calibration of the reflected intensity as well as the round-trip time between the antenna and the MLS under investigation. Only then the MLS can be measured (see Fig. 2.2c, d – blue curve). To compensate the noise of the source and switch a TRM illumination configuration must be modified, to enable simultaneous measurement and calibration. This can be achieved by making a double arm illumination configuration which is out of scope of this paper.

Experimental Results

Here we show some results that we got from practical measurements by means of TRM for three and single layer structures. First, we consider a three-layer structure

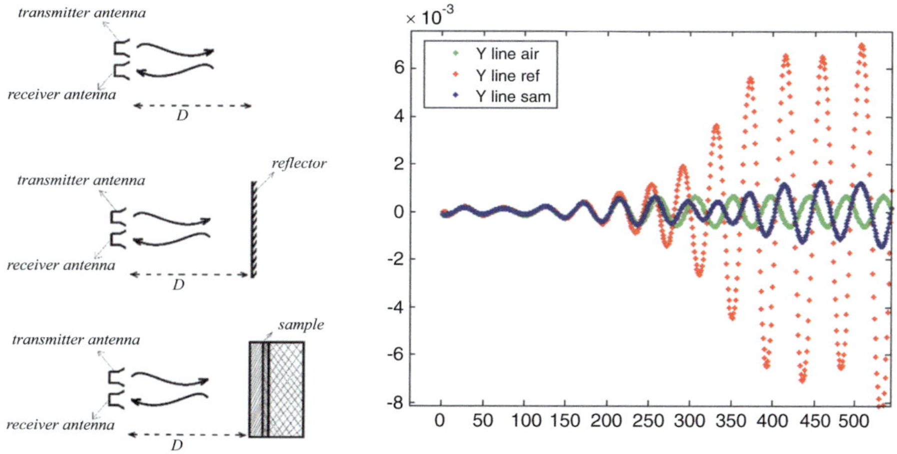

Fig. 2.2 (**a**, **b** and **c**) TRM calibration and measurement procedure. (**d**) In this figure, we correspond Air, Ref and Sam on each other for finding nose

to demonstrate the extraction of geometrical and complex permittivity data for each with a single frequency. In this measurement, we used two PVC sheets which have 500 × 300 × 50.04 mm and 500 × 300 × 20.08 sizes, respectively. In the middle of them we had a gap of 30.11 mm. Look at Table 2.1. The single layer structure under consideration is a PVC-slab with a size of 500 × 300 × 103 mm. Data of the layers can be found in Table 2.2.

2.3 Graphene Based Metal Waveguide Modulators

In our work, we focused on *EM wave* propagation in the rectangular waveguide that shown in Fig. 2.3; the rectangular waveguide croess-section is situated in the XY plane, and is composed-of two layers divided by a monolayer of graphene: the upper part of the waveguide cross-section ($0 < y < d_b$) is accopied by the first (buffer) layer with relative dielectric permittivity ε_b, and the lower part ($-d_a < y < 0$) is occupied by the second (air) layer with relative dielectric permittivity ε_a. The waveguide walls are assumed to be perfectly conducting metals, meaning that its relative dielectric permittivity modulus $|\varepsilon_m| \to \infty$, this condition is well satisfied for metals at *sub-THz* frequencies. The graphene layer is supposed to be isolated from the vertical metal walls.

The standard electrodynamic boundary conditions for the electric field tangential component E_t and magnetic field normal component H_n are used at the perfectly conducting metal walls: $E_t = H_n = 0$. At the surface $y = 0$, coincident with the monolayer of graphene, the boundary conditions are as follows: the continuity condition of the electric field tangential component and a modified boundary

Table 2.1 Experimental results of thicknesses and complex permittivity for three-layer structure which is included of PVC, air and PVC

Three-layer structure	First layer	Second layer	Third layer
Relative permittivity (NRW)	2.88 ± 0.06 − 0.11j ± 0.06j	1.00 + 0.00j (air)	2.88 ± 0.06 − 0.11j ± 0.06j
Relative permittivity (TRM)	2.81 ± 0.05 − 0.14j ± 0.05j	1.13 ± 0.08 − .04j ± 0.08j	2.71 ± 0.10 − 0.10j ± 0.10j
Thickness (mm) (Caliper)	50.04 ± 0.01	30.11 ± 0.01	20.08 ± 0.01
Thickness (mm) (TRM)	52.71 ± 0.06	37.21 ± 0.07	26.07 ± 0.1

Table 2.2 Complex permittivity and thickness of single layer structure: TRM versus Caliper and NRW

	Thickness(mm) (first layer)	Permittivity(ε_r)
Caliper and NRW data	103 ± 0.1	2.38 ± 0.06 − 0.05j ± 0.06j
TRM data	106.3 ± 0.1	2.54 ± 0.05 − 0.15j ± 0.05j

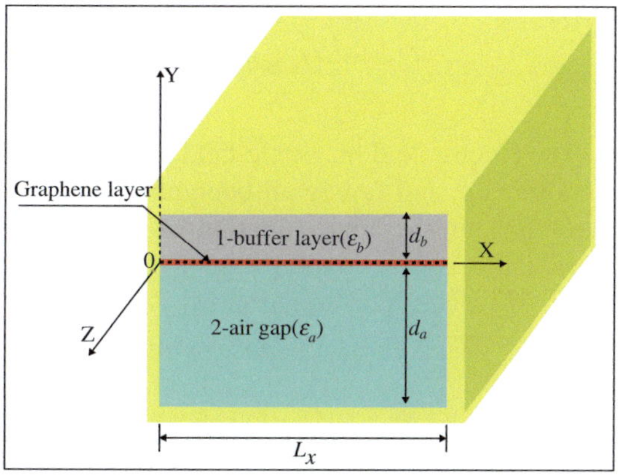

Fig. 2.3 Schematic view of the considered rectangular waveguide structures

conditions due to the presence of the graphene layer for the magnetic field tangential components. As the graphene layer can be characterized by a surface electron concentration n_s and a surface conductivity σ, the latter boundary conditions can be written down as follows:

$$H_{x,1}(y=0) - H_{x,2}(y=0) = -\sigma E_z(y=0) \tag{2.1a}$$

$$H_{z,1}(y=0) - H_{z,2}(y=0) = -\sigma E_x(y=0) \tag{2.1b}$$

Where $H_{j,1}$ and $H_{j,2}$ are the tangential components of the magnetic field amplitude in buffer and air layers, respectively, $E_j(y=0)$ is the amplitude of the electric field tangential component at y = 0, $(j = x, z)$.

In general the graphene surface conductivity is given by:

$$\sigma(\omega) = \sigma^{\text{intra}}(\omega) + \sigma^{\text{inter}}(\omega) \tag{2.2a}$$

At room temperature, intraband transitions give the largest contribution to the optical conductivity of graphene for low terahertz frequencies $f < 30\ THz$ and Fermi levels $E_F > 0.05\ eV$. The intraband graphene conductivity is given as:

$$\sigma^{\text{intra}}(\omega) = i\frac{2Te^2 \ln(2\cosh(E_F/2T))}{\pi\hbar^2\omega(1 + i/\omega\tau)} \tag{2.2b}$$

Where ω is waveguide mode angular frequency, e is the electron charge, E_F is the Fermi energy level of graphene, τ is electron relaxation time, $T = k_B t$ are the graphene electron energy, Fermi energy and temperature (k_B is Boltzmann constant and t is temperature in Kelvin) [17].

For graphene (the Fermi velocity $\upsilon_F = 10^6 m/s$ $_s$s), the mobility value is suggested to be equal to 10^4 $cm^2/V.s$ [18].

The dispersion equation for rectangular waveguide structure is given as follows:

$$Q_2 + \left(Q_1 \cot(Q_1 d_b) - i\frac{\sigma(\omega, q)\omega}{c^2 \varepsilon_0} \right) \tan(Q_2 d_a) = 0 \tag{2.3}$$

The electric and magnetic field components of *TE* modes do not depend on the x coordinate, and satisfy all boundary conditions of the rectangular layered waveguide only in the approximation of perfectly conducting metal waveguide walls. Beyond this approximation when the metal dielectric permittivity is supposed to be finite. The dispersion equation for *TE* modes of planar metal layered waveguide – without vertical metal walls at $x = (0, L_x)$. Which is given as:

$$\begin{aligned} &\left(1 + \frac{Q_2}{Q_m}\right)\left(Q_2 + \beta Q_1 - \frac{\sigma k_0}{c\varepsilon_0}\right) \\ &+ \exp(-2iQ_2 d_a)\left(1 - \frac{Q_2}{Q_m}\right)\left(Q_2 - \beta Q_1 + \frac{\sigma k_0}{c\varepsilon_0}\right) \\ &= 0, \end{aligned} \tag{2.4}$$

$$\text{here } \beta = \frac{1 + (iQ_1 \tan(Q_1 d_b))/Q_m}{i \tan(Q_1 d_b) + Q_1/Q_m}, \; Q_m = \sqrt{k_0^2 \varepsilon_m - q^2}, \; Q_1 = \sqrt{k_a^2 \varepsilon_b - \left(\frac{\pi n}{L_x}\right)^2 - q^2},$$

$Q_2 = \sqrt{k_a^2 \varepsilon_a - \left(\frac{\pi n}{L_x}\right)^2 - q^2}$, $k_0 = \frac{\omega}{c}$, q is the waveguide mode propagation constant, $c = \frac{1}{\sqrt{\varepsilon_0 \mu_0}}$ is the light speed in vacuum, ε_0 and are the permittivity and permeability of vacuum. In the approximation of perfectly conducting metal waveguide walls, (2.4) reduces to (2.3). Solutions of dispersion (2.3) and *TE* modes field distributions for planar metal waveguides are almost coincident with *TE* modes of rectangular metal waveguide if $|\varepsilon_m| \gg 1$, although *TE* modes field distributions for planar metal waveguide do not strictly satisfy accurate boundary conditions at the vertical metal waveguide walls. For more details see [15, 16].

In more detail the structure of the *TE* modes of the rectangular waveguide, and planer, described by dispersion (2.3) and (2.4). All *TE* modes of the rectangular waveguide are labeled as TE_{0m} modes where subscript index *0* means that these modes are the solutions for $n = 0$ and the index m identifies the solution number; the modes in the absence of graphene layer are denoted as TE_{0m}^0 modes. Solutions of

(2.3) are given by the well-known dispersion relations for modes of a hollow rectangular waveguide in the absence of buffer layer when $d_b = 0$:

$$q^2 = k_0^2\varepsilon_a - \frac{m^2\pi^2}{d_a^2}, \tag{2.5}$$

where $m = 1,2,3,\ldots$

Numerical calculations were performed for the waveguide structures of Fig. 2.3 with the following material parameters: air relative dielectric permittivity $\varepsilon_a = 1$, the size of the rectangular waveguide is 864 × 432 μm at $f = 0.3\ THz$, for buffer layer we consider SiO_2 with $\varepsilon_b 4 + 0.04i$ valid for frequencies $f < 3\ THz$ [19]. For Au waveguide wall metal $\varepsilon_m = (-0.112 + 2.427i) \times 10^6$ [20]. Therefore, dispersion relations for TE_{0m} modes of the planar and rectangular waveguides are very close to each other.

The dependences of the first TE waveguide mode propagation constant versus Fermi level for planar metal waveguide are shown in (Figs. 2.4a, b) at frequency $f = 0.3\ THz$ and different buffer thicknesses. It is clear that modulation is strongest for buffer thickness $d_2 = 150\ \mu m$ which approximately coincides with the mode cutoff for $E_F = 0.5\ eV$). It is clear that both real and imaginary parts of the propagation constant can be effectively modulated by changing the Fermi level in graphene which lead to phase and amplitude modulation, respectively.

The propagation constant dependences of Fig. 2.5a belong to the TE_{01} mode for Fermi level $E_F < 0.15\ eV$, to the TE_{02} mode for $0.15\ eV \le E_F \le 0.22\ eV$, to the TE_{03} mode for $0.22 eV \le E_F \le 0.28\ eV$, to the TE_{04} mode for $0.28\ eV \le E_F \le 0.35\ eV$ and to the TE_{05} mode for $E_F \ge 0.35\ eV$. The propagation constant dependences of waveguide modes are represented by long-range modes for buffer thickness $d_b = 200 \mu m$ if $E_F < 0.14\ eV$, for buffer thickness $d_b = 250 \mu m$ if $E_F < 0.18\ eV$ and within the whole range of $E_F \le 0.4\ eV$ for buffer thickness $d_b = 300 \mu m$. The modulation of both real and imaginary parts of the mode propagation constant is strong for these waveguide modes; for example, the waveguide modes being long-range ones for buffer thicknesses $d_b = (200, 250) \mu m$ without graphene layer are converted into short-range ones increasing the Fermi level up to $0.2\ eV$.

The propagation constant dependences of Fig. 2.5b belong to the TE_{02} mode for Fermi level $E_F < 0.15\ eV$ and to TE_{01} mode for $E_F \ge 0.15\ eV$; these dependences are represented by long-range modes for buffer thicknesses $d_b = (200,\ 250\ \text{and}\ 300) \mu$ m within the whole range of Fermi levels $E_F \le 0.4\ eV$. In this figure the propagation constant modulation is the largest for buffer thickness $d_b = 175 \mu m$ and Fermi level; the maximum of TE_{01} mode propagation constant imaginary part for reaches the value about $5\ cm^{-1}$ for – this value is about 25 times larger compared to the same value without graphene layer.

It is clear that both real and imaginary parts of the propagation constant can be effectively modulated by changing the Fermi level in graphene which lead to phase and amplitude modulation, respectively.

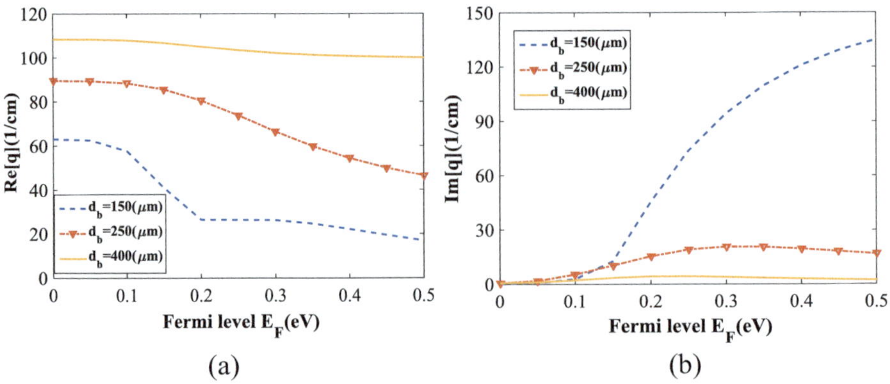

Fig. 2.4 Dispersion dependences of the first *TE* waveguide mode propagation constant versus Fermi level for f = 0.3 THz, for $d_2 = (150, 250, 400)$ μm, for planar metal waveguide (**a**) the real part, (**b**) the imaginary part

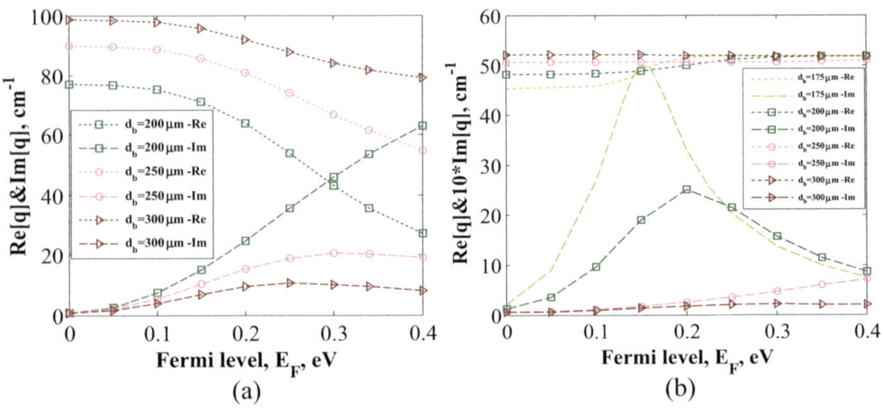

Fig. 2.5 (**a**) Dependences of TE_{01} modes propagation constants versus Fermi level for different buffer thickness d_b; real parts (*Re[q]*), and imaginary parts multiplied by 10 (*Im[q]*), for $f = 0.3\ THz$. (**b**) Dependences of waveguide modes propagation constants q versus Fermi level for different buffer thickness; real parts (*Re[q]*), and imaginary parts (*Im[q]*), for $f = 0.3\ THz$, for $f = 0.3\ THz$. (**c**) Dependences of waveguide modes propagation constants q versus Fermi level for different buffer thickness; real parts (Re[q]), and imaginary parts multiplied by 10 (Im[q]), for f = 0.3 THz

2.4 Conclusion

The novel blind algorithm for extraction of layer thickness and dielectric properties of each layer in a multi-layered structure is demonstrated, which can be used for nondestructive testing in many industrial applications. The transient radar is validated by comparing obtained results with a well-known method such as NRW. The experimental results of the transient radar are shown for single and three- layer structures.

The supported modes by a rectangular waveguide with an integrated graphene structure have been presented in the approximation of perfectly conducting waveguide metal walls. It is shown that propagation constants real and imaginary parts can be considerably modulated by varying the concentration of graphene layer electrons especially near the regions of the modes modification; this leads to the possibility of considerable modulation of transmitted and reflected electromagnetic waves in metal waveguides with integrated graphene structure.

Acknowledgments The authors acknowledge the Vrije Universiteit Brussel (VUB) through the SRP-project M3D2, IMEC through the BISENS project, the COST-action MP1204, the NATO support for the workshop THz Diagnostics of CBRN effects and Detection of Explosives & CBRN, and the Libyan Ministry of Higher Education for technical and financial support.

References

1. J. R. Little, Jr., “High-resolution, nondestructive imaging of dielectric materials,” U.S. Patent 2010 0 283 483 A1, Nov 11, 2010.
2. J. R. Little, Jr., “Apparatus and method for nondestructive testing of dielectric materials,” U.S. Patent 6 359 446 B1, Mar 19, 2002.
3. A. M. Nicolson and G. F. Ross, “Measurement of the intrinsic properties of materials by time-domain techniques,” IEEE Trans. Instrum. Meas., vol. 19, no. 4, pp. 377–382, Nov. 1970.
4. W. B. Weir, “Automatic measurement of complex dielectric constant and permeability at microwave frequencies,” Proc. IEEE, vol. 62, no. 1, pp. 33–36, Jan. 1974.
5. W. Huashen, J. Shan, W. Guodong, and X. Ke, “Electromagnetic parameters test system based on A refined NRW transmission/reflection algorithm,” in Proc Int Symp Microw, Antenna, Propag EMC Technol Wireless Commun, Hangzhou, China, Jun 2007, pp. 1276–1280.
6. A. L. de Paula, M. C. Rezende, and J. J. Barroso, “Modified Nicolson–Ross–Weir (NRW) method to retrieve the constitutive parameters of low-loss materials,” in Proc SBMO/IEEE MTT-S Int Microw Optoelectron Conf, Natal, Brazil, Oct 2011, pp. 488–492.
7. S. P. Luttrell and C. J. Oliver, “Prior knowledge in synthetic aperture radar (SAR) processing,” in Proc Roy Signals Radar Establishment, Malvern, UK, Jan. 1985, pp. 73–77.
8. R. J. Stakenborghs, “Validation of a microwave based inspection system for HDPE butt fusions,” in Proc ASME, Conf Pressure Vessels Piping, vol 5 Anaheim, CA, USA, Jul. 2014, pp. 1–11.
9. L. Udpa and S. S. Udpa, “Neural networks for the classification of nondestructive evaluation signals,” IEE Proc F (Radar Signal Process), vol. 138, no. 1, pp. 41–45, Feb. 1991.
10. T. Nagatsuma, H. Nishii, and T. Ikeo, “Terahertz imaging based on optical coherence tomography,” Photon. Res., vol. 2, no. 4, pp. B64–B69, Aug. 2014.
11. A. Pourkazemi; J. H. Stiens; M. Becquaert; M. Vandewal, “Transient Radar Method: Novel Illumination and Blind Electromagnetic/Geometrical Parameter Extraction Technique for Multilayer Structures,” in IEEE Transactions on Microwave Theory and Techniques, vol. PP, no.99, pp.1–14, doi:10.1109/TMTT.2017.2665633
12. K. Novoselov, et al, “Electric field effect in atomically thin carbofilts,” Science, 2004, 306, 666–669.G. T. Reed, Silicon Photonics (Wiley, 2008).
13. E. Carrasco et al, “Reflectarray antenna at terahertz using graphene,” IEEE Antennas Wireless Propag. Lett. 12, 253–256, 2013.
14. H. Alkorre, et al, “Coupled TM plasmon-polaritons features of graphenemetal layered structure in the sub-THz frequency range,” J. Opt., 17 045003, 2015.

15. G. Shkerdin, et al, "Modified TM & TE waveguide modes and reflectivity by graphene layer in coupled-graphene-metal multilayer structure in sub-Terahertz frequency," J. Opt., 17, 05500620117, 055006, 2015.
16. G. Shkerdin, et al, "Modified TE modes of metal waveguide with integrated graphene structure in the sub-terahertz frequency range," IET, MAP. 0365.R2. 1751–8725, 2015.
17. L. A. Falkovsky, "Optical properties of graphene and IV-VI semiconductors Phys. Usp., 51, 887–897, 2008.
18. M. Jablan, et al. "Plasmonics in graphene at infrared frequencies". Phys. Rev. B, 80, 245435, 2009.
19. R. Kitamura, et al, "Optical constants of silica glass from extreme ultraviolet to far infrared at near room temperature," Appl. Opt, 46 8118–33, 2007.
20. A. Ordal, et al, "Optical properties of fourteen metals in the infrared and far infrared: Al, Co, Cu, Au, Fe, Pb, Mo, Ni, Pd, Pt, Ag, Ti, V, and W," Appl. Opt., 24, 4493–4499, 1985.

Chapter 3
Near-Field Characterization of Conductive Micro-resonators for Terahertz Sensing

Irina Khromova and Oleg Mitrofanov

Abstract Near-field (NF) terahertz (THz) time-domain spectroscopy (TDS) is an excellent tool for direct studies of THz electromagnetic resonances occurring on a micrometre scale. Micro-resonators are at the heart of numerous promising THz sensing and detecting solutions. Experimental studies of individual micrometre-scale THz resonances are essential, yet extremely challenging for the common far-field spectroscopic methods due to extreme sensitivity requirements. NF THz spectroscopy and microscopy are non-contact techniques for spectroscopic studies of individual micro-resonators and mapping the field patterns of THz resonant modes excited in individual conductive or insulating micro-objects. They give access to essential parameters of micro-resonators, including their resonance frequency, local field enhancement and quality factors. It allows for material and structural characterisation of micro-objects. Using the example of carbon micro-fibres, we show the advantages of NF THz TDS for non-contact THz conductivity probing and direct observation of the fundamental and the third-order surface-plasmon resonance modes in conductive THz micro-resonators.

Keywords Terahertz spectroscopy • Metamaterials • Carbon fibre • Near field

3.1 Introduction

To advance terahertz (THz) technology for security applications, including the detection of substances through their distinctive THz and mid-infrared 'signatures' [1], it is essential to develop efficient THz sensors. Extreme sensitivity can be achieved in resonant systems. Micrometre-sized objects, made of conductive materials, exhibit resonant response at THz frequencies. They can be used as individual

---

I. Khromova (✉)
Department of EEE, University College London, London, UK
e-mail: i.khromova@ucl.ac.uk

O. Mitrofanov
Department of EEE, University College London, London, UK

CINT, Sandia National Laboratories, Livermore, California, USA

M.F. Pereira, O. Shulika (eds.), *THz for CBRN and Explosives Detection and Diagnosis*, NATO Science for Peace and Security Series B: Physics and Biophysics,
DOI 10.1007/978-94-024-1093-8_3

sensors or macro-scale sensing devices, based on metamaterials – artificial structures made of pre-designed resonant 'meta-atoms'.

It is essential to experimentally study and characterise individual micro-resonators. However, common far-field THz spectroscopic techniques cannot access the micrometre-scale resonances due to low sensitivity. We successfully used the near-field THz time-domain spectroscopy (NF THz TDS) for direct experimental characterisation of individual conductive micro-resonators [2].

Through NF THz TDS, we show that at THz frequencies, carbon micro-fibres (CMFs) exhibit pronounced surface-plasmon resonances. The NF THz TDS allows for non-contact conductive material characterisation in micrometre-sized objects, crucial for the practical development of resonance-based devices for the emerging THz technology for security applications.

3.2 NF THz TDS for Individual Micro-objects

In our experiments, individual micro-resonators (CMFs) are attached to a thin low-density polyethylene (LDPE) film pulled over a sample-holder frame. They are excited by a linearly-polarised broadband THz pulse (0.5–2.5 THz) generated in a ZnTe nonlinear crystal through optical excitation. The experiments were performed at 22 ± 1 °C. The enhanced near-field induced in the micro-resonators is detected by photoconductive antenna covered by a metallic screen with a subwavelength aperture [3, 4]. We use two different probes with 5 and 10 μm square aperture. Moving the time-delay stage (Fig. 3.1), we perform the near-field spectroscopy in the time-domain.

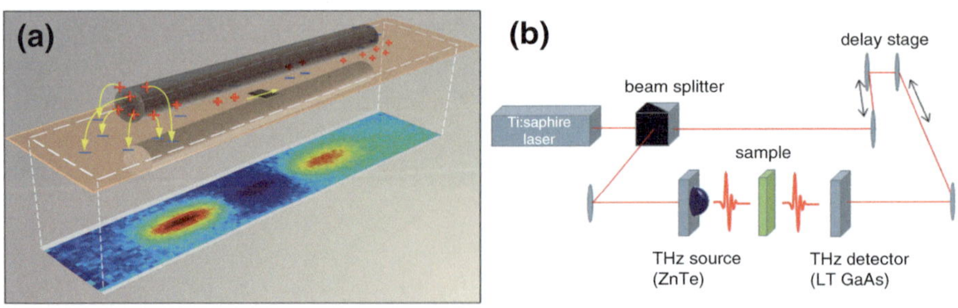

Fig. 3.1 (**a**) Scheme of the NF THz TDS setup; (**b**) Schematic image of a CMF probed by the aperture in the metallic plane of the probe. The *arrows* represent the electric field lines of the complex sample-probe mode. The coloured map is the detected enhanced field distribution measured at a fixed time delay by scanning the sample with respect to the aperture

3.3 Surface-Plasmon Resonances in THz Micro-resonators

Carbon micro-fibres (CMFs) are milled and chemically-treated carbon fibres of essentially graphitic structure produced and supplied by STC of Applied Nanotechnologies. Their diameters are 7 μm and their lengths vary between 30 and 300 μm. Depending on the fabrication procedure, the DC resistivity of similar unmilled carbon fibers is of the order of 10 μΩ × m [5] and strongly depends on temperature [6]. The plasma frequency of graphite in the conductive plane lies significantly above the THz region [7, 8] and corresponds to carrier density of $\approx 10^{18}$ cm^3.

Given their high conductivity and dimensions comparable to half-wavelength of the working THz range, CMFs can be considered as half-wave dipole resonators at THz frequencies [9, 10].

We performed NF THz TDS on CMFs of different lengths, and detected surface-plasmon resonances (dimensional resonances of induced current) of different modal orders [2, 11]. In the employed setup, the detected signal is proportional to the gradient dE_z/dx of the field induced between the conductive CMF and the detector metallic plane of the THz detector as shown in Fig. 3.1a [2, 12]. As a consequence, when mapping the field of the fundamental resonance mode of a CMF, the maximum signal is measured at the centre of the CMF rather than at its edges.

Figure 3.2 shows the maps of the detected THz field obtained through raster scanning the sample with respect to the probe aperture. Figure 3.2a is the map of the detected signal at its highest peak (fixed time delay at 1.65 ps after the arrival of the THz pulse) for a 133 μm-long CMF. This image corresponds to the fundamental surface-plasmon resonance mode of the CMF – half-a-cycle oscillation of the induced current propagating along the structure. Figure 3.2b,c map the signal detected behind a 207 μm-long CMF at 1 ps and 1.27 ps after the arrival of the THz pulse. They reflect the detection of a third-order surface-plasmon resonance mode – three halves of the induced current oscillation along the CMFs' length.

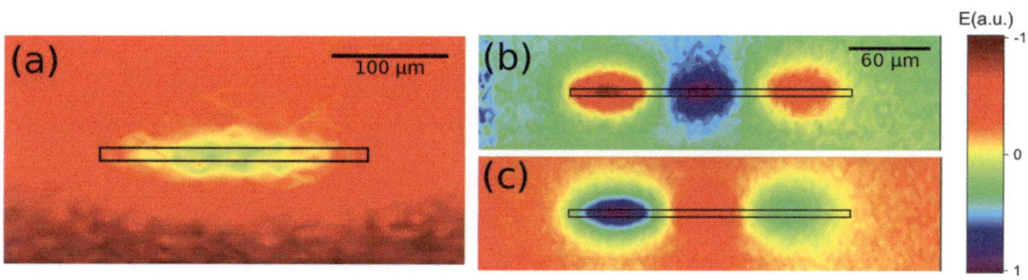

Fig. 3.2 (**a**) fundamental and (**b,c**) third-order surface-plasmon resonance detected in the near-field of a 133 μm-long and 207 μm-long CMF, respectively. The maps were acquired at (**a**) 1.65 ps, (**b**) 1 ps, and (**c**) 1.27 ps after the arrival of the incident THz pulse

3.4 Non-contact Conductivity Probing via NF THz TDS

While it is possible to get a rough estimate of the average conductivity of a CMF composite via far-field spectroscopic methods, directly measuring the conductivity of a single CMF is a non-trivial task. Through NF THz TDS, we were able to probe the material properties of individual CMFs in a non-contact way.

To access this information about the CMFs' constituent material, we studied the peak amplitude of the enhanced near-field as well resonance frequencies of the fundamental surface-plasmon resonance mode of the CMFs. Both of the two parameters are highly influenced by both the material properties of the resonator and the electromagnetic properties of the environment, including the substrate and the metallic probe effects.

We used full-wave numerical simulations in CST Microwave Studio to predict and separate these effects. The model included the realistic physical parameters of the experimental THz set-up, including the near-field aperture-based THz probe.

Figure 3.3 shows the coloured maps of the enhancement factor (amplitude peak of the detected field normalised to the incident field probed by the THz antenna) and the resonance frequency of the fundamental surface-plasmon resonance. Each colour band represents numerical data obtained for a certain band of CMF conductivity values. Plotting the experimental data against such maps allowed us to estimate the conductivity value corresponding to each of the individual CMFs. We found that the conductivity of the CMFs is $10^4 - 10^5$ S/m. As the conductivity increases, the areas become narrower. Thus, the proposed method works best for 'weakly' conductive materials, such as doped semiconductors and semimetals.

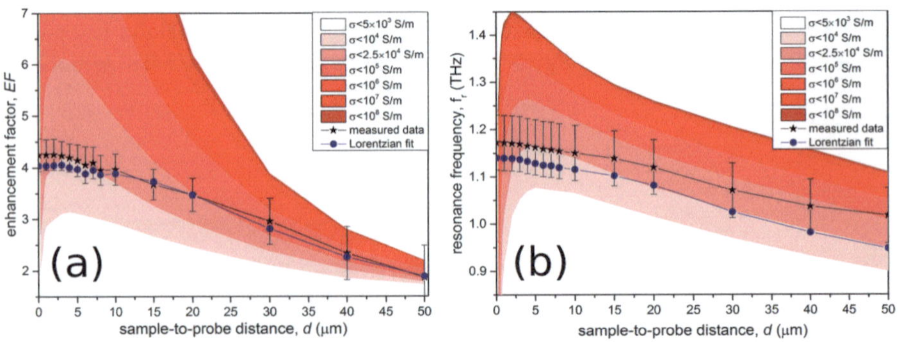

Fig. 3.3 (**a**) Enhancement factor EF and (**b**) resonance frequency f_r of an 80 μm-long CMF plotted against the sample-to-probe distance. *Coloured zones mark* numerically calculated areas corresponding to different conductivity bands within the range from 10^3 S/m (material with skin depth comparable to the diameter of the CMFs) to 10^8 S/m (metals). *Black curves* and *symbols* represent the measured *EF* and f_r, while the *blue curves* and *symbols* correspond to *EF* and f_r extracted from Lorentzian fitted experimental spectra. The *error bars* reflect the effects of noise and the spectral resolution of 58 GHz (Reproduced from [2], with the permission of AIP Publishing)

3.5 Substrate-Induced Resonance Frequency Shift in CMFs

The resonance frequency of the surface-plasmon resonances in CMFs are extremely sensitive to the dielectric properties of the surrounding material: we observed how the 12.5 μm-thick LDPE substrate with dielectric permittivity $\varepsilon_{sub} = 2.39$ shifted the resonance frequencies of the fundamental and third-order modes in different CMFs by a factor of 0.89 as compared to the numerically predicted values for equal, but free-standing CMFs (Fig. 3.4).

The substrate-induced frequency shift is due to electric field localisation in the immediate vicinity of the CMFs and can be estimated as a factor of $\sqrt{2/(1+\varepsilon_{LDPE})}$. Such behaviour can be utilised for sensing applications, in particular, for probing the dielectric permittivity of substances on a micrometre scale.

3.6 Conclusion

Studying THz resonant micro-objects in the near-field provides a deep insight into the electromagnetic properties of individual resonators functioning on versatile physical principles. NF THz TDS is an excellent experimental platform for detecting and observing the excitation and evolution of surface-plasmon resonance modes in conductive micro-resonators. We applied this technique to CMFs of different length and observed the excitation of the fundamental as well as the third-order resonance mode.

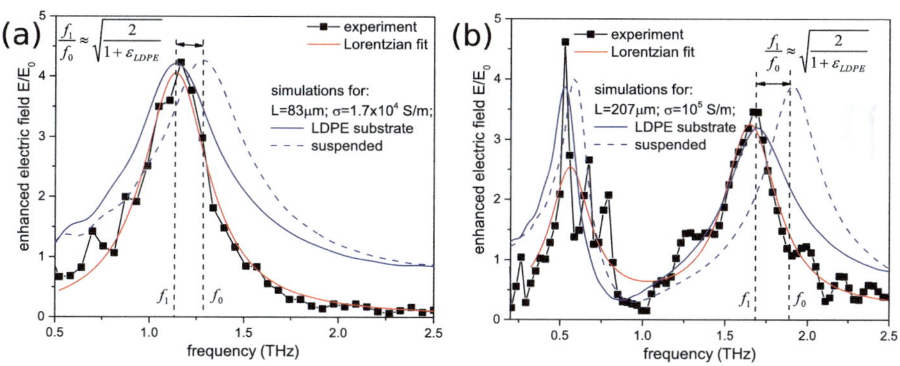

Fig. 3.4 Experimentally obtained enhanced field spectra (*black* curve and *squares*) for CMFs placed at 3 μm from the THz probe: (**a**) length L = 83 μm; full-wave numerical simulations performed for $\sigma = 1.7 \times 10^4$ S/m; (**b**) L = 207 μm; full-wave numerical simulations performed for $\sigma = 10^5$ S/m. The *red* curves are the Lorentzian fits. The full-wave numerical simulations were performed for CMFs on LDPE substrates (solid *blue* curve) and suspended CMF (dashed *blue* curve). A spectral shift of the resonance frequency is observed due to the change in the permittivity of the sample's environment

We demonstrate that NF THz TDS is a promising non-contact method for probing the electric conductivity of individual conductive resonators. The method works best for 'weakly' conductive materials as compared to metals: we successfully measured the conductivity of $10^4 - 10^5$ S/m in individual CMFs. It is extremely difficult to perform analogous experiments in the far field using conventional THz TDS systems: assuming their typical beam spot size is approximately 1 mm in diameter, the presence of a 100 μm-long CMF would only manifest itself in a mere $\approx$1.5% intensity dip in transmitted field at the resonance frequency.

We also show that the near-field response of conductive micro-resonators is highly dependent on the environment. We found that the resonance peak of the fundamental and the third-order surface-plasmon resonance mode in conductive CMFs is shifted by the polyethylene substrate by a factor of 0.89, confirming the sensitivity of CMFs towards the dielectric properties of surrounding materials. This finding can be useful for THz sensing applications.

References

1. M.F. Pereira, TERA-MIR radiation: materials, generation, detection and applications. Opt. Quant. Electron. 46, 491–493 (2014) DOI: 10.1007/s11082-014-9883-5
2. I. Khromova, M. Navarro-Cía, I. Brener, J. L. Reno, A. Ponomarev and O. Mitrofanov, Dipolar resonances in conductive carbon micro-fibers probed by near-field terahertz spectroscopy, Appl. Phys. Lett. 107 (2), 021102 (2015) DOI: 10.1063/1.4926628
3. R. Knab, A. J. L. Adam, E. Shaner, H. J. A. J. Starmans, and P. C. M. Planken, Terahertz near-field spectroscopy of filled subwavelength sized apertures in thin metal films, Opt. Express 21, 1101 (2013) DOI: 10.1364/OE.21.001101
4. A. J. Macfaden, J. L. Reno, I. Brener, and O. Mitrofanov, 3 μm aperture probes for near-field terahertz transmission microscopy. Appl. Phys. Lett. 104 (1), 011110 (2014) DOI: 10.1063/1.4861621
5. D. D. Chung, Carbon Fiber Composites (Butterworth-Heinemann, Boston, 1994)
6. I. L. Spain, K. J. Volin, H. A. Goldberg, and I. L. Kalnin, Unusual electrical resistivity behavior of carbon fibers, Solid State Commun. 45, 817 (1983) DOI: 10.1016/0038-1098(83)90807-4
7. R. C. Tatar and S. Rabii, Electronic properties of graphite: A unified theoretical study, Phys. Rev. B, 25, 4126 (1982) DOI: 10.1103/PhysRevB.25.4126
8. L. G. Johnson and G. Dresselhaus, Optical Properies of Graphite, Phys. Rev. B, 7, 2275 (1973) DOI: 10.1103/PhysRevB.7.2275
9. S. Schelkunoff and H. Friis, Antennas: Theory and Practice, Applied mathematics series (Wiley, 1952)
10. E. S. Barnard, J. S. White, A. Chandran, and M. L. Brongersma, Spectral properties of plasmonic resonator antennas, Opt. Express, 16, 16529 (2008) DOI: 10.1364/OE.16.016529
11. O. Mitrofanov, I. Khromova, Th. Siday. R. J. Thompson, A. N. Ponomarev, I. Brener, J. L. Reno, Near-Field Spectroscopy and Imaging of Subwavelength Plasmonic Terahertz Resonators. IEEE Trans THz Science and Technol. 99, 1–7 (2016) DOI: 10.1109/TTHZ.2016.2549367
12. R. Mueckstein, C. Graham, C. C. Renaud, A. J. Seeds, J. A. Harrington, and O. Mitrofanov, Imaging and Analysis of THz Surface Plasmon Polariton Waves with the Integrated Sub-wavelength Aperture Probe, J. Infrared Millimeter Terahertz Waves 32, 1031 (2011) DOI: 10.1007/s10762-011-9811-8

Chapter 4
Bio-chemical Process Monitoring with Terahertz Sensor

Y. Zhang, V. Matvejev, S. Declerck, S. Stroobants, G. Pandey, G. He, D. Mangelings, D. Maes, S. Muyldermans, and J. Stiens

Abstract An in-house developed Label-free Immobilization-free Terahertz (THz) sensor has shown its potential in various applications [1]. In this paper, we present the experimental results in the field of High Performance Liquid Chromatography (HPLC) and protein crystallization. Ultra-Violet (UV) detectors are very commonly used as a detection technique in HPLC systems. However, it relies on labelling when target substances do not absorb UV light. Therefore, it has the drawback of increased cost and time consumption. Our experimental result shows that THz sensor has the potential to replace UV sensor integrated on the HPLC machine as it is able to detect both UV and non-UV absorbing substances without labelling. Meanwhile, the THz sensor is also deployed for the first time to follow a protein crystallization process.

Keywords Millimeter wave • HPLC • Protein crystallization • THz • Dielectric permittivity

---

Y. Zhang • V. Matvejev • G. Pandey • G. He • J. Stiens (✉)
Laboratory of Micro- and Photoelectronics, LAMI-ETRO, Vrije Universiteit Brussel (VUB), Pleinlaan 2, Brussels 1050, Belgium
e-mail: jstiens@etro.vub.ac.be

S. Declerck • D. Mangelings
Department of Analytical Chemistry and Pharmaceutical Technology, Center for Pharmaceutical Research (CePhaR), Vrije Universiteit Brussel-VUB, Laarbeeklaan 103, B-1090 Brussels, Belgium

S. Stroobants • D. Maes
Structural Biology Brussels, Vrije Universiteit Brussel, Pleinlaan 2, 1050 Brussels, Belgium

S. Muyldermans
Cellular and Molecular Immunology, Vrije Universiteit Brussel, Pleinlaan 2, 1050, Brussel, Belgium

M.F. Pereira, O. Shulika (eds.), *THz for CBRN and Explosives Detection and Diagnosis*, NATO Science for Peace and Security Series B: Physics and Biophysics,
DOI 10.1007/978-94-024-1093-8_4

4.1 Introduction

In biological solutions, the presence of bio-molecules alters the hydrogen bond network dynamics of water molecules in its vicinity. These molecules, also known as hydration shell, exhibit different dielectric permittivity than normal bulk water molecules in the THz frequency range (30 GHz–1 THz) [2]. The amount of hydration shell water molecules in liquid is closely related to the amount of bio-molecules, their shape, conformation flexibility and the distribution of hydrophobic and hydrophilic surface. In other words, the change in concentration or conformation of these bio-molecules will translate to dielectric permittivity change in the THz frequency range. In previous work [3], a THz sensor operating @ 270 GHz has been made to probe permittivity changes in aqueous solutions. It is a label-free, immobilization-free real-time sensor that features very high sensitivity. In this study, another sensor topology operating @ 60 GHz is developed for the same purpose. High performance liquid chromatography (HPLC) and protein crystallization measurements are chosen as proof-of-concept studies to validate the sensor design.

4.2 THz Sensor & Setup

The THz sensor developed in house consists of a metal waveguide section, and a capillary tube as liquid fixture (shown in Fig. 4.1a). THz wave is launched from one side of the waveguide. The interaction between incident wave and liquid results in reflection and transmission. If the impedances of the sensor and the rest of the THz set-up are properly matched, very low reflection (S_{11} Parameter) can be achieved at a certain frequency. When the dielectric permittivity of the liquid changes as a result of bio-molecular interactions, the wave impedance of the sensor will also deviate from its original value. The mismatch of impedance between the THz sensor and the other parts creates a larger reflection signal. In real measurements, a reference liquid is first injected into the waveguide via capillary tube. Then impedance tuning is applied to achieve desired S_{11} signal level (blue curve in Fig. 4.1b). When bio-molecular interactions occur (increased concentration of certain substance, protein interaction etc.), the change of the liquid's dielectric permittivity leads to an increased S_{11} Parameter (red curve in Fig. 4.1b). This difference will be registered for further analysis.

The measurement set-up consists of (Fig. 4.2a):

- Vector network analyzer that produces and detects 60GHz signal
- Directional coupler that separates reflected wave from the forward propagating wave
- THz sensor that is very sensitive to dielectric change of liquid under test.

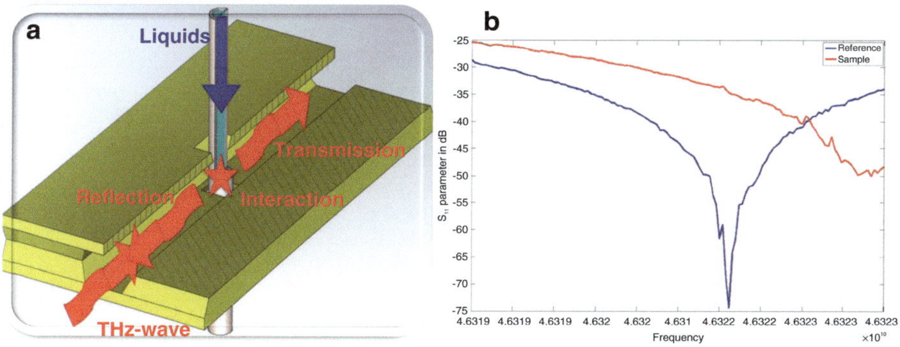

Fig. 4.1 (**a**) THz sensor internal structure (**b**) different responses of reference liquid and sample liquid

Such measurement set-up has shown its capability in the previous work [3]. When compared to other THz sensor designs by bench marking water-ethanol contrast measurements (Fig. 4.2b), it shows superior sensitivity. Note that the THz sensor used in this paper is denoted as A1 in the graph which is 30 dB more sensitive than the closest K [4].

4.2.1 High Performance Liquid Chromatography System Setup & Measurements

HPLC system is often used to separate individual ingredients in a mixture, and measure the concentration separately. Such system typically consists of (Fig. 4.3):

- HPLC pump(s) to keep mobile phase flowing through the entire system
- An injector to inject samples onto the mobile phase
- A stationary phase to separate individual substances from the sample injected
- Detector to register chromatograph in order to measure the concentration of each component in the sample. There exists various techniques [15]. In our case, UV detector is integrated on the available machine.

To avoid extra complexity, the stationary phase is left out since the target of this study is detection of substances instead of separation. As a result, liquid with only one target substance can be injected each time. The THz sensor coupled HPLC system is shown in Fig. 4.4.

Three substances as candidates have been chosen for this proof-of-concept study. Two UV-absorbing substances are: Trans-Stilbene-Oxide (TSO) and Praziquantel (PRA). One non-UV-absorbing substance is: Sorbitol. Since TSO and PRA only dissolve poorly in water, acetonitrile (ACN) is chosen as mobile

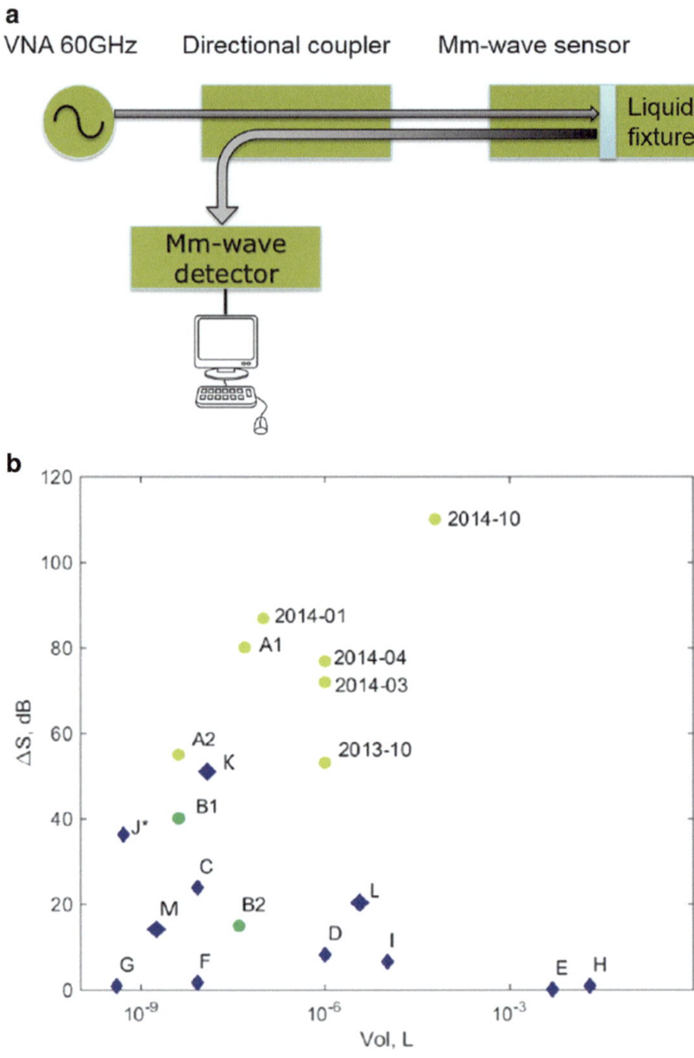

Fig. 4.2 (**a**) THz sensor measurement set-up. (**b**) Bench marking performance comparison of different THz sensors. *A1* used in this paper. *A2*, *B1*, *B2* are designs made in the past. The other *yellow dots* are theoretical designs. *Blue diamonds* are sensors presented in literature (J[5], C[6], M [7], L[8], D[9], I[10], H[11], E[12], F[13], G[14])

phase in such case. Water is chosen to be the mobile phase for the measurement with sorbitol for the same reason.

During the measurements, impedance tuning is applied when only mobile phase (ACN or water) is in the THz sensor. A continuous-wave measurement on this frequency is started at the moment of injection. Since the distance between the THz sensor and the UV detector is fixed, we expect that the time delay between the two

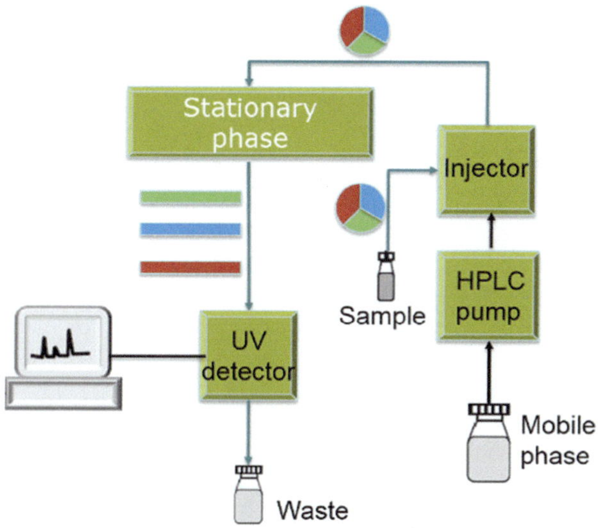

Fig. 4.3 High performance liquid chromatography system (*Red*, *green* and *blue colors* depict different substances in a mixture)

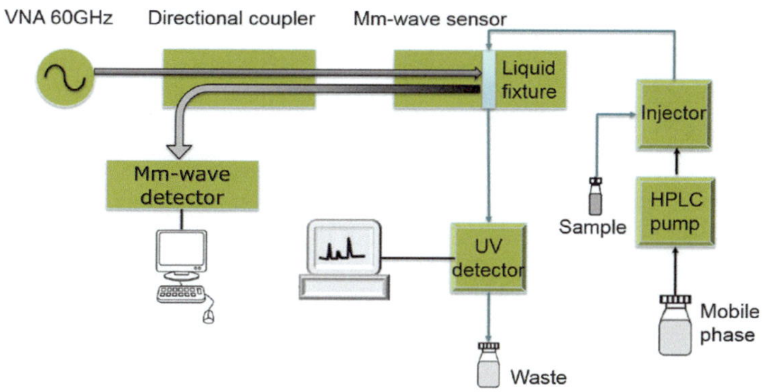

Fig. 4.4 THz sensor coupled HPLC system

sensor responses is also constant. A series of measurements has been conducted and the results are shown in Fig. 4.5.

As shown in Fig. 4.5a, very little signal variation can be observed from both sensors with a blank injection. For THz sensor, since the impedance of the entire system is matched with ACN, the injected ACN does not introduce any permittivity change. The UV detector has a flat line response since ACN is not UV absorbing. As PRA and TSO are injected (Fig. 4.5b, c), both detectors respond to the substances. The time delay also stays constant around 1.3 min. Since sorbitol is non-UV absorbing, the UV detector was not able to detect it (Fig. 4.5d). However, the THz sensor gave a very obvious response.

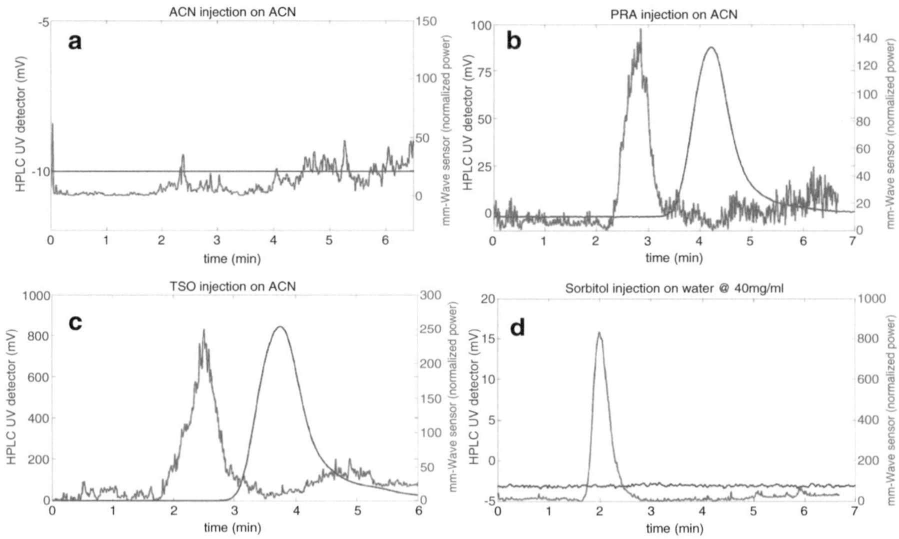

Fig. 4.5 (**a**) Blank injection: ACN injection on mobile phase ACN. (**b**) 10 mg/ml PRA injection on mobile phase ACN. (**c**) 10 mg/ml TSO injection on mobile phase ACN (**d**) 40 mg/ml Sorbitol injection on mobile phase water

4.2.2 Protein Crystallization Measurements

Protein crystallization plays a vital role in understanding the structure of proteins [16]. Though it has been extensively studied, the mechanism of this process is still not clear. UV-VIS and Dynamic light scattering are very commonly used to follow this process among various other techniques. In this study, the capability of THz sensor in this field will be tested.

The THz set-up is slightly modified to measure protein crystallization. The liquid to be measured will be extracted from an eppendorf tube into the sensor structure (Fig. 4.6a). The driving force will be provided by a 25ul syringe.

When the concentrations of lysozyme (40, 50 &60 mg/ml) and NaCl (40 mg/ml) are properly chosen in a mixture(0.1 M Sodium Acetate as buffer), crystallization process can be established at an predictable speed. Increasing the concentration of either lysozyme or NaCl will leads to reduction of time needed for such process. In each measurement, the freshly made crystalizing mixture is chosen as reference liquid. The sensor system is tuned so that very low S_{11} parameter is achieved. This reference is then compared to the same process at different times (Fig. 4.6b).

As shown in Fig. 4.6c, with increasing lysozyme concentration, the time needed to observe a rise in the THz signal shortens. This corresponds to the accelerated crystallization process.

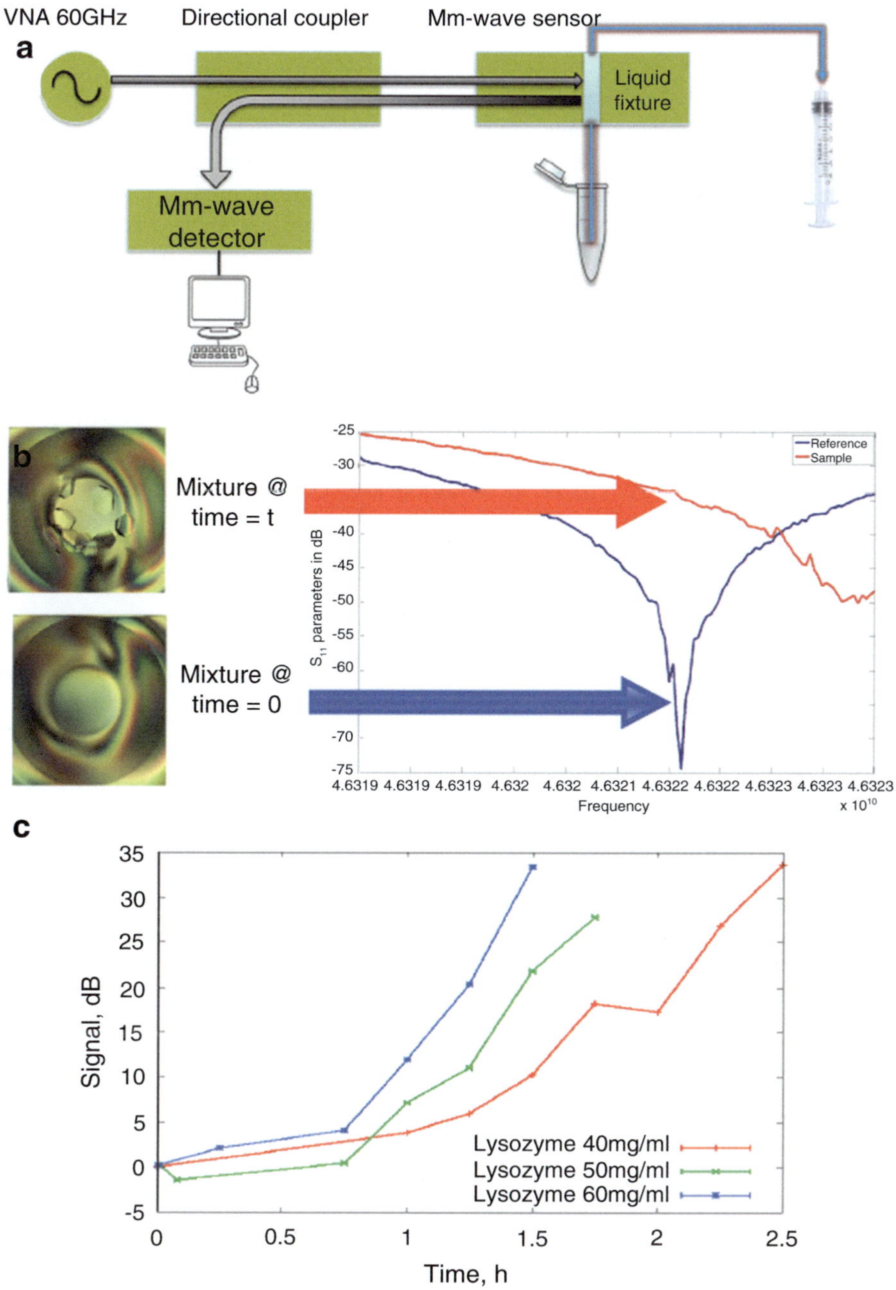

Fig. 4.6 (**a**) Crystallization THz sensor setup (**b**) Illustration of reference liquid and measured sample (**c**) Crystallization measurement results with different lysozyme concentrations

4.3 Conclusion

Although further studies are still required to validate the THz sensor for the two applications demonstrated above, through our proof-of-concept study, it is already clear that this sensor can track protein crystallization processes. The label-free feature also gives it an advantage over the UV detector in the field of High-performance Liquid Chromatography. In the near future, the focus will be shifted to quantitative measurements of HPLC so that the concentration of injected substances can be determined from measurement curves. For protein crystallization, efforts need to be made to increase the time resolution of the measurements.

Acknowledgments The authors acknowledge the Vrije Universiteit Brussel (VUB) through the SRP-project M3D2, the FWO-Vlaanderen through FWOAL682 "Building blocks of lab-on-chip system for label-free monitoring of bio-molecular interactions" and the FWOAL611 "Millimeter wave sensor solutions for chromatographic analyzer systems of today and tomorrow", the, the COST-action MP1204, TERAMIR the NATO support for the Workshop on THz Diagnostics of CBRN effects and Detection of Explosives & CBRN.

References

1. V. Matvejev, et al. Integrated waveguide structure for highly sensitive THz spectroscopy of nano-liter liquids in capillary tubes, Progress in Electromagnetic Research, 121, 89–101 (2011)
2. Leitner, D.M., et al. Solvation dynamics of biomolecules: modeling and terahertz experiments, HFSP J., 2008, 2, (6), pp. 314–323
3. V. Matvejev, Y. Zhang, J. Stiens, High performance integrated THz sensor for detection of biomolecular processes in solution, IET Microwaves, Antennas & propagation, 8(6), 394–400 (2014)
4. Weina Liu, A Novel Technology for Measurements of Dielectric Properties of Extremely Small Volumes of Liquids, International Journal of Antennas and Propagation, Volume 2016, Article ID 1436798, 5 pages
5. Cui Y., Y. He, P. Wang. "A Quadrature-Based Tunable Radio-Frequency Sensor for the Detection and Analysis of Aqueous Solutions" *IEEE MICROWAVE AND WIRELESS COMPONENTS LETTERS*, vol. 24, pp. 490–492, 2014.
6. R. Mendis, V. Astley, J. Liu, D.M. Mittleman, Appl. Phys. Lett 95, 171113 (2009)
7. S. M. Hanham et al. APPLIED PHYSICS LETTERS 107, 032903 (2015)
8. Vaishali Rawat et al., Journal of Applied Physics 116, 164106 (2014)
9. M. Neshat et al., Elect Lett 44(17), 1020 (2008)
10. L. Wendling, et al. Real Time Monitoring and Detection of Alcohol Using Microwave Sensor Technology, IEEE Proc. Conf. Developm eSystems, 2009
11. N. Wiwatcharagoses, et al, Michigan State Uni report 2009
12. A. Sklavounos, N. Barker, IRMMWTHz-IEEE Proc, p2 (2011)
13. Laurette, S, et al. Subterahertz characterization of ethanol hydration layers by microfluidic system, Applied Physics Letters, 97, 111904 (2010), DOI:http://dx.doi.org/10.1063/1.3488832
14. T. Chretiennot et al. A Microwave and microfluidic planar resonator for efficient and accurate complex permittivity characterization of aqueous solutions. IEEE Trans Microw Theory Techn. 2013 Feb;61(2):972–978
15. W.W. Christie. Detectors for HPLC of lipids with special reference to evaporative lightscattering detection. In Advances in Lipid Methodology – One. Edited by W.W. Christie, Oily Press, Ayr. pp. 1992. 239–271
16. N. E. Chayen, "Turning protein crystallisation from an art into a science," Current Opinion in Structural Biology, vol. 14, pp. 577–583, Oct. 2004

Chapter 5
Submm-Wave Reflection and Transmission Technique for Testing and Monitoring of Biochemical Solutions

A.I. Gubin and S.A. Vitusevich

Abstract The investigation of dielectric permittivity of liquids in a wide frequency range is important not only for liquid identification, but these studies provide very useful information on liquid under test as well. We show that the method of investigation of the complex conductivity of superconducting samples by grazing incidence reflectivity measurements can be also used for investigation of dielectric properties of different biochemical solutions. The optimal incident angle was obtained by calculation of the reflection and transmission coefficients at the liquid-dielectric interfaces for given conditions. It is determined as the maximum of the reflection coefficient sensitivity to the liquid permittivity changes. The measurement cell based on Teflon prism allowing grazing incidence angles on liquid media is designed. The experimental results of the transmission coefficient through the prism filled with water and air demonstrates that the technique can be used for the liquid investigation in a wide frequency range. The other advantage of the method is the possibility of measurement of liquids with different thicknesses and especially samples with unknown thickness. This technique can be also applied for determination of explosives and monitoring of different biochemical solutions.

Keywords Grazing incidence reflectivity • Submillimeter waves • Liquid investigation

---

A.I. Gubin (✉)
Department of Solid-State Radiophysics, O. Ya. Usikov Institute for Radiophysics and Electronics NAS of Ukraine, Kharkiv 61085, Ukraine
e-mail: gubin@ire.kharkov.ua

S.A. Vitusevich
Peter Grünberg Institute (PGI-8), Forschungszentrum Jülich, Jülich 52425, Germany
e-mail: s.vitusevich@fz-juelich.de

M.F. Pereira, O. Shulika (eds.), *THz for CBRN and Explosives Detection and Diagnosis*, NATO Science for Peace and Security Series B: Physics and Biophysics,
DOI 10.1007/978-94-024-1093-8_5

5.1 Introduction

The investigation of the complex conductivity (or permittivity) of superconducting samples by grazing incidence microwave reflectivity measurements is a promising method for studying of the samples with different thicknesses in a wide temperature range (i.e. samples with temperature dependent conductivity) [1, 2]. The main feature of the method is usė of grazing incident p-polarized wave in order to increase the sensitivity of the measurement, which opens prospects for such a kind of material investigation. The sensitivity of the reflection coefficient to changes of permittivity increases up to the Brewster angle, which is almost 90° for metals and superconductors. Besides it was shown the applicability of the technique to investigate other materials such as material with giant magnetoresistance [3]. In this paper, we show that the method can be also useful for the investigation of different biochemical solutions by measuring of reflection or\and transmission coefficients of the samples under test. Knowing the liquid properties in a wide frequency range is important for liquid identification as well as for providing additional useful information about liquid under test.

5.2 Calculation Results

First, we calculate the Brewster angles for different interfaces, such as vacuum-liquid, Teflon-liquid and sapphire-liquid. The values are important for a new approach development.

Brewster angle for interface of the two media with refraction indices n_1 and n_2 can be calculated by:

$$\theta_B = \tan^{-1}\left(\frac{{n_2}^2}{n_1}\right).$$

The refraction indices we obtain from the permittivity by $n = \sqrt{\varepsilon}$ (we suppose $\mu = 1$). The dielectric permittivity of liquids can be obtained by known expressions [4]. Brewster angle is lower for higher permittivity of the first media n_1. For the frequency of 300 GHz the angle is equal to 72°, 65° and 44° for vacuum-water, Teflon-water and sapphire-water interfaces, respectively. The angles are not near to 90° (as in the case of vacuum-metal interface) and can be in principle realized experimentally.

It is convenient to use plane wave approach to determine the reflection and transmission coefficient for dielectric-liquid interface [1]. The final reflection and transmission coefficients through the layered media were obtained using Fresnel equations and Snell's law for p-polarized wave (p means that the vector $\boldsymbol{E}$ lies in the plane of the incidence). The reflection and transmission coefficients through the Teflon-water and Teflon-ethanol interfaces as a function of incidence angle at

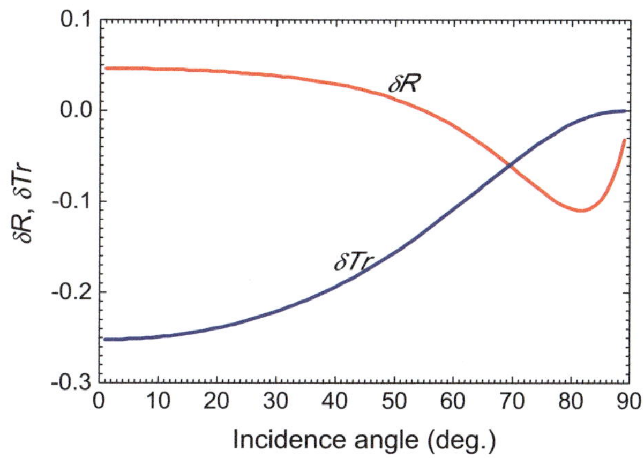

Fig. 5.1 *δR* and *δTr* dependencies on incident angle

frequency of 300 GHz were obtained. To determine the sensitivity of reflection and transmission coefficients to changes of permittivity it is convenient to introduce value of $\delta R(\theta) = R_{water}(\theta) - R_{ethanol}(\theta)$, which is related to the sensitivity of reflection and transmission coefficients to difference in permittivity of ethanol and water.

The value of δT decreases with increase of incidence angle up to 90° (Fig. 5.1). The value of δR demonstrates nonmonotonous behavior. It decreases with increase of incidence angle and reaches 0 at angle about 55°. At higher angles of incidence the value of δR increases and reaches the value more than two times higher than the value at normal incidence. The results reflect that for given conditions (investigations by reflection of the *p*-polarized wave incident from Teflon media to liquid media) the studies at incident angle equal to 82° have to be most sensitive to the liquid permittivity changes and the angle of 82 is the most optimal incident angle.

5.3 Experimental Results

To measure the reflection and transmission coefficients in submm wave range the experimental setup, consisting of backward-wave oscillator (BWO), Teflon lenses, parabolic mirrors, wire grid polarizes, measurement cell holder, optical chopper, attenuator and Gollay-cell detector, was used (Fig. 5.2). The setup has a number of the BWO's, which cover 80–700 GHz frequency range.

The signal generated by BWO is radiated by horn and focused by Teflon lens. The optical chopper is used for modulation of the incident wave at the frequency 15 Hz, which is optimal for the Gollay-cell detector operatation. The attenuator allows variation of the incident power in the range from 1 to 30% and adjusting it value depending on different measured samples and cells under test. The tilted wire grid polarizers reduce standing wave by reflection of the waves with other polarization to free space. The parabolic mirror focuses the beam to the measurement

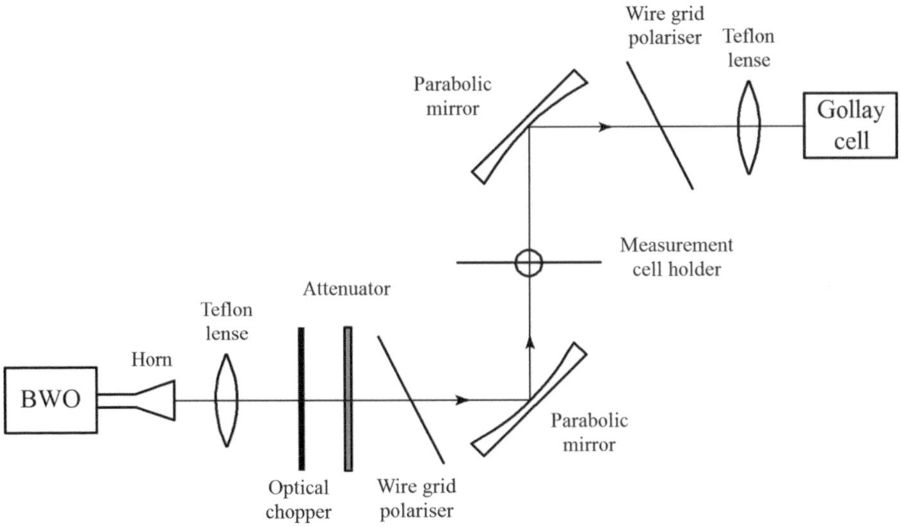

Fig. 5.2 Schematics of the measurement setup

cell with liquid, which is placed in the cell holder (Fig. 5.2). The diameter of the beam spot at the surface of the cell is about 6 mm.

The holder was covered with absorbing material to reduce the reflections. It allows also rotating the measurement cell with liquid around the axis and reach arbitrary incidence angle. The horn, the wire grid polarizers and the rotated measurement cell holder are oriented in such a way in order to reach the p-polarization of the incident wave, i.e. to place vector E into the incident plane. After the measurement cell the signal passes to the parabolic mirror and to the tilted wire grid polarizer. Teflon lens focus the incident power into the window of Gollay cell detector. The signal detected by Gollay cell is amplified by lock-in amplifier with optical chopper reference frequency.

The simplest way to design the measurement cell is to place the liquid between two dielectric layers, for example between two sapphire substrates. However, in this case the incident angle to measurement cell and the incident angle to liquid layer are different. Calculation of the propagation angles in the three layer cell with two layers of sapphire shows the changes of incident angle on sapphire-liquid interface from 0° to only 20° with changing incident angle on cell from 0° to 90°, correspondingly. These angles are far from optimal incidence angle. On the other hand, this peculiarity can be used to reach optimal grazing incident angles (for example for Teflon-liquid interface the angle is equal to 82°).

The wave incident onto the surface of 45° Teflon prism at the 45° angle will propagate in Teflon at 30°, which will result in 75° angle of incidence at Teflon-liquid interface (Fig. 5.3). At such an angle of incidence the sufficient increase of the reflection sensitivity to the dielectric permittivity changes is predicted (Fig. 5.1).

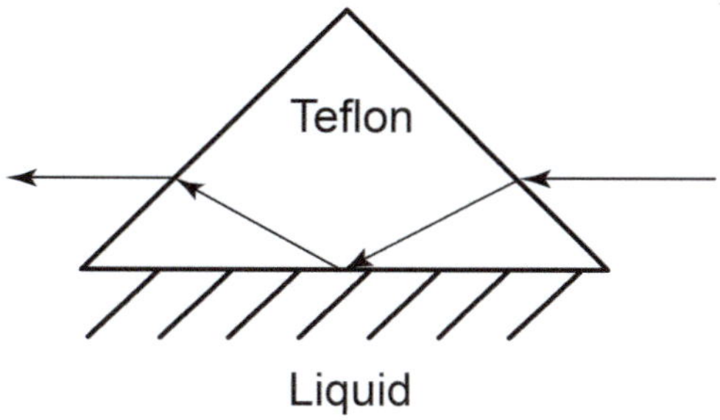

Fig. 5.3 Propagation of the wave in 45° Teflon prism, realizing 75° angle of incidence onto the liquid media

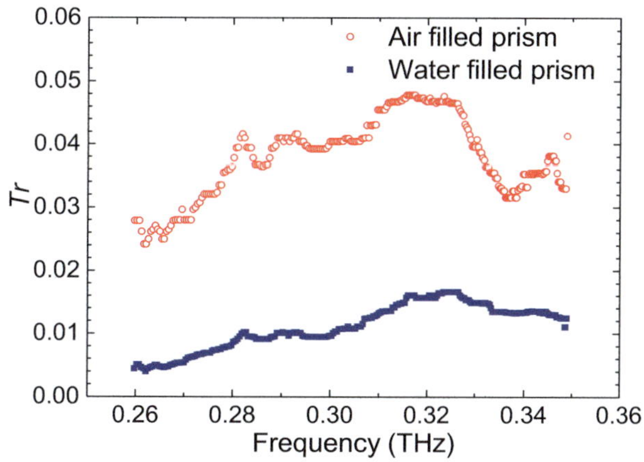

Fig. 5.4 The transmission through the prism filled with the air and water dependence on the frequency

The measured transmission *Tr* through the prism (i.e. reflection from the liquid media) dependence on frequency is shown in Fig. 5.4. There is strong difference between the air and liquid filling curves. The results prove the feasibility of new approach for liquid investigation under grazing incidence reflectivity conditions. It should be noted, that the difference in reflectivity of the liquid filled and air filled three layered measurement cell under normal incidence conditions is at the level of the measurement accuracy (not shown here).

The other advantage is the possibility of measurement of liquids with different thicknesses and especially the samples with unknown thicknesses. It should be noted, that the usage of the prism is necessary only in the case of liquids measurements. In the case of solids (for example solid explosive materials), it is enough to tilt the plane of the sample and measure the reflection coefficient.

5.4 Conclusion

The transmission and reflection coefficient dependencies on angle for three layered dielectric-liquid-dielectric measurement cell with ethanol and water fillings on incidence angle were calculated. The optimal incidence angle for water investigation was obtained. The measurement setup for measurements of reflection and transmission coefficients was developed and calibrated. The measurement cell was realized using the Teflon prism providing the grazing incidence at 75° onto the liquid media. The measured and calculated data clearly demonstrate the possibility of utilization the grazing incident reflectivity method for liquid as well as solid materials.

Acknowledgments Authors are grateful to Sergiy Pud for useful discussions and help with the measurements.

References

1. N. T. Cherpak, A. I.Gubin, A. A. Lavrinovich, Microwave Reflectivity of HTS Film - Dielectric Substrate Structure at Arbitrary Incidence Angles. Telecommunications and Radio Engineering 55, 81–90 (2001). doi: 10.1615/TelecomRadEng.v55.i3.120
2. A.I. Gubin, N.T. Cherpak, A.A. Lavrinovich, MM Wave Conductivity Measurement Technique by Grazing Incidence Reflectivity. EuMC'2008, October 2008, Amsterdam, The Netherlands, Symposium proceeding, 1074–1076 (2008). doi: 10.1109/EUMC.2008.4751643
3. B.I. Belevtsev, N.T. Cherpak, I.N. Chukanova, A.I. Gubin, V.B. Krasovitsky and A.A. Lavrinovich, Direct-current transport properties of and microwave absorption in a bulk ceramic sample and a film of $La_{0.5}Sr_{0.5}CoO_{3-\delta}$: magnetic inhomogeneity effects. J. Phys.: Condens. Matter 14, 2591–2603 (2002). doi: 10.1088/0953-8984/14/10/310
4. S. Mashimo, T. Umehara, Structures of water and primary alcohol studied by microwave dielectric analyses. J. Chem. Phys. 95, 6257–6260 (1991). doi: 10.1063/1.461546

Chapter 6
Radiation Induced Molecular Damage Addressed by Terahertz Spectroscopy – A Theoretical Study

M. Mernea, O. Calborean, I. Vasile, S. Avram, and D.F. Mihailescu

Abstract Nuclear radiation can seriously damage biological samples. Some radiation effects are the production of reactive radicals resulted from water radiolysis and random bond breakings, protein fragmentation, cross-linking or aggregation resulted from random ionizations in proteins. Similar effects were described in the case of X-rays irradiation. Therefore, we assumed that not only nuclear radiation, but also the lower energy X-rays produce flexibility changes in proteins that can be addressed by Terahertz (THz) spectroscopy. Here we simulated and compared the THz spectra of two *Torpedo californica* acetylcholinesterase structures that were exposed to synchrotron radiation (hard X-ray of ~13 keV) over different periods of time. Our results show a clear difference between simulated absorptions, which allows us to conclude that radiation induced changes in proteins structures and flexibility should be visible on THz spectra. By detecting even subtle effects produced by radiation softer than nuclear radiation, THz spectroscopy should be useful in detecting effects of higher energy radiation that is part of the chemical, biological, radiological and nuclear (CBRN) threat.

Keywords THz spectroscopy • X-ray damage • Spectra simulation

6.1 Introduction

Due to its high energy, nuclear radiation has the potential to produce serious molecular level damage to biological matter. Among the different types of nuclear radiation, γ-ray impact on the structure of proteins has been extensively investigated. In this case, indirect and direct effects were described. The indirect

M. Mernea • O. Calborean • S. Avram • D.F. Mihailescu (✉)
Department of Anatomy, Animal Physiology and Biophysics, Faculty of Biology, University of Bucharest, Bucharest, Romania
e-mail: dan.mihailescu@bio.unibuc.ro

I. Vasile
Horia Hulubei National Institute for Physics and Nuclear Engineering, Magurele, Romania

M.F. Pereira, O. Shulika (eds.), *THz for CBRN and Explosives Detection and Diagnosis*, NATO Science for Peace and Security Series B: Physics and Biophysics,
DOI 10.1007/978-94-024-1093-8_6

effect is mediated by the reactive radicals (OH^- and H^+) resulted from water radiolysis that diffuse trough the solution and interact with specific solvent accessible chemical groups from proteins [9]. The magnitude of indirectly occurred radiation damage depends on the temperature, as radicals diffusion is impaired at significantly low temperatures [10]. The direct radiation effect is due to random ionizations in protein structures that result in random bond breakings which cause protein fragmentation, cross-linking or aggregation [4].

Similar direct and indirect effects can be seen when irradiating proteins with X-rays. Such effects have been intensively researched since synchrotron X-rays applied on protein crystals are used for solving protein structures. In X-ray crystallography, radiation damage is limited by cryocooling, but damage still occurs when the exposure time increases [6]. Primary damage produced by the uneven photoelectric absorption consists of bonds breakage and formation of electron, hydroxyl and hydrogen radicals. Secondary damage is produced by these radicals that break additional bonds and produce even more radicals [7].

Several studies conducted on proteins such as *Torpedo californica* acetylcholinesterase (*Tc*AChE), hen egg white lysozyme [13, 14], chymotrypsin inhibitor [13], thaumatin, elastaze, insulin, lysozyme, ribonuclease A or trypsin [5] revealed that X-ray exposure causes breakage of disulfide bonds and a loss of definition for some residues side chains due to the loss of carboxyl groups from aspartate and glutamate residues, hydroxyl groups from tyrosine and methylthio groups from methionine. Not all residues are equally susceptible [14], the residues from enzymes active sites being particularly sensitive [14]. Increased values of atoms B factors (temperature factors, a measure of isotropic displacements of atoms in the crystal) are signs of radiation damage [13, 14].

It was shown that radicals formation is accelerated by low pH values [2]. A transient drop of pH value was identified upon water radiolysis [1]. Molecular simulations showed that the "acidic spike" reaches the lowest values of ~3.3 in spherical spurs and ~2.5 in cylindrical tracks during the first ns after irradiation. After ~1 μs to ~0.1 ms, the pH returns at the value of 7 [8].

The present study is based on our expectations that not only the high energy nuclear radiation that causes protein fragmentation, but also lower energy radiation such as hard X-rays, changes the flexibility of proteins and that such changes should be visible on Terahertz (THz) spectra. Here we simulated and compared the THz spectra of two *Tc*AChE structures that were exposed to synchrotron radiation (hard X-ray of ~13 keV) over different periods of time [14]. The changes of pH due to water radiolysis were also taken into account.

6.2 Materials and Methods

6.2.1 *Modelling of Irradiated Proteins*

Here we used the two *Tc*AChE structures under the Protein Data Bank codes 1QID and 1QIG [14]. These structures were determined in an experiment during which nine data sets were collected on the same crystal, each data set being collected over 19 minutes. The structures that we chose were determined based on the first (1QID – will be further called point A) and fourth (1QIG – will be further called point D) data sets.

When building the models of irradiated *Tc*AChE we considered the breakage of Cys254-Cys265 disulphide bond in point D structure and the protonation state of both structures induced by the changes in water pH upon radiolysis. The protonation states of the structures at 3 pH values (2, 4, and 6) were assigned using PROPKA [12]. Resulting total charges of the structures were 61 e at pH 2, 40 e at pH 4 and 6 e at pH 6. Modelling and optimization of structures was performed with CHARMM software [3].

6.2.2 *THz Spectra Simulation*

THz spectra of point A and point D structures were simulated using normal modes analysis (NMA). Required energy minimizations and NMA calculation was performed with CHARMM software [3]. Based on the results, normal modes intensities and the broadening of intensity lines was performed as described in our previous work [11].

6.3 Results and Discussions

In our study we considered two *Tc*AChE structures (*point A* and *D*) that were chosen by taking into account the following aspects: (i) *point A* structure presents the smallest amount of radiation damage, having an average B factor of 30.5 $Å^2$; (ii) in comparison to point A, *point D* structure presents clear signs of radiation damage, namely an average B factor of 41.1 $Å^2$ and the breakage of Cys254-Cys265 disulphide bond; (iii) the two structures have a similar number of crystal water molecules (333 in the case of point A and 335 in the case of point D) [14]. These properties mapped on the two structures are presented in Fig. 6.1a, b. Besides the presented differences, the two proteins present a similar conformation, as the root mean squared deviation (RMSD) calculated between them is 0.13 Å. *Point A* and *point D* structures were protonated according to three pH

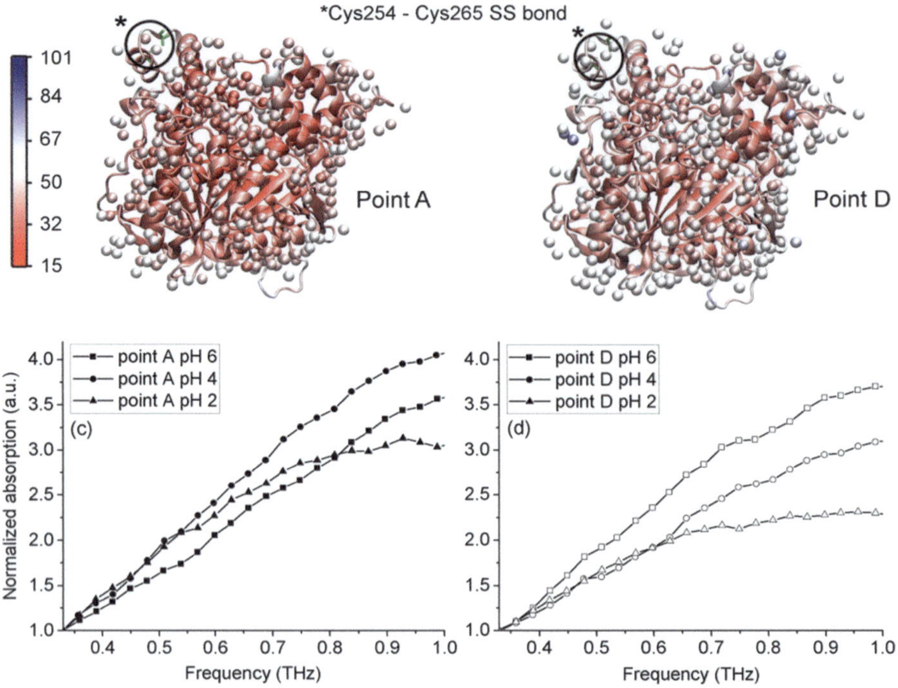

Fig. 6.1 (**a**) *Tc*AChE *point A* (PDB code 1QID [14]) and (**b**) *Tc*AChE *point D* (PDB code 1QIG [14]) structures. The backbones of the proteins are represented as cartoon and the oxygen atoms from the crystal water molecules are represented by their van der Waals spheres. The structures are coloured according to their B factors and the used colour scale is given in the *left*. The location of the disulphide (SS) bond that is broken in point D structure is *circled* on the figures. (**c**), (**d**) Simulated THz spectra of *Tc*AChE *point A* (**c**) and *point D* (**d**) structures protonated according to the pH values of 2, 4 and 6

values in order to account for the pH drop and subsequent recovery that occurs upon water radiolysis.

THz spectra of protonated *point A* and *D* structures were simulated in the frequency range comprised between 0.33 and 1 THz. For an easier comparison, the spectra were normalized to the absorption at 0.33 THz. Resulting data presented in Fig. 6.1c, d shows that, overall, the absorptions of considered structures are different. The absorptions increase linearly with frequency in the 0.33–0.7 THz. The slopes of the lines that fit the spectra are presented in Fig. 6.2. In the case of *Tc*AChE *point A* structure, the structure protonated according to pH 6 presents a lower line slope than the structures at pH 4 and pH 2, which in turn present similar line slopes values. In the case of *Tc*AChE point D, the structure at pH 6 presents a higher slope value than the structures at pH 4 and pH 6, which are slightly different. At each pH value, we notice a clear difference between the slopes of the lines that fit the theoretical absorption of *point A* and *point D* structures.

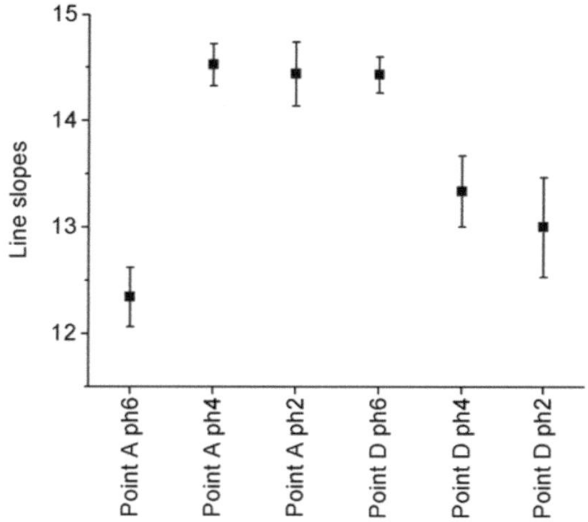

Fig. 6.2 Slopes of the lines that fit the simulated *Tc*AChE *point A* and *D* structures in the 0.33–0.72 THz frequency range

6.4 Conclusion

The results presented here show a clear difference between the simulated THz absorption of the irradiated *Tc*AChE structures protonated according to each pH value. Therefore, we conclude that THz spectroscopy should be able to identify the radiation damage produced by X-rays, a type of radiation with lower energy than nuclear radiation and that radiation damage produced by higher energy radiation should be even more straightforward to identify using THz spectroscopy.

Acknowledgments Research was financially supported by the Romanian National Authority for Scientific Research through the UEFISCDI PNII Grants PCCA-198/2014, PCCA-89/2012, IDEAS 137/2011 and HIVERA 4-004/2013.

References

1. M. Anbar, J.K. Thomas, Pulse Radiolysis Studies of Aqueous Sodium Chloride Solutions1. J. Phys. Chem. 68(12), 3829–3835 (1964). doi:10.1021/j100794a050.
2. K. Bobrowski, J. Holcman, J. Poznanski, K.L. Wierzchowski, Pulse radiolysis studies of intramolecular electron transfer in model peptides and proteins. 7. Trp–>TyrO radical transformation in hen egg-white lysozyme. Effects of pH, temperature, Trp62 oxidation and inhibitor binding. Biophys Chem 63(2–3), 153–166 (1997). doi:S0301-4622(96)02226-0 [pii].
3. B.R. Brooks, C.L. Brooks, 3rd, A.D. Mackerell, Jr., L. Nilsson, R.J. Petrella, B. Roux, Y. Won, G. Archontis, C. Bartels, S. Boresch, A. Caflisch, L. Caves, Q. Cui, A.R. Dinner, M. Feig, S. Fischer, J. Gao, M. Hodoscek, W. Im, K. Kuczera, T. Lazaridis, J. Ma, V. Ovchinnikov, E. Paci, R.W. Pastor, C.B. Post, J.Z. Pu, M. Schaefer, B. Tidor, R.M. Venable, H.L. Woodcock, X. Wu, W. Yang, D.M. York, M. Karplus, CHARMM: the biomolecular simulation program. J. Comput. Chem. 30(10), 1545–1614 (2009). doi:10.1002/jcc.21287.
4. W.M. Garrison, Reaction mechanisms in the radiolysis of peptides, polypeptides, and proteins. Chemical Reviews 87(2), 381–398 (1987). doi:10.1021/cr00078a006.

5. M. Gerstel, C.M. Deane, E.F. Garman, Identifying and quantifying radiation damage at the atomic level. J. Synchrotron. Radiat. 22, 201–212 (2015). doi:10.1107/S1600577515002131.
6. A. Gonzalez, C. Nave, Radiation-Damage in Protein Crystals at Low-Temperature. Acta Crystallographica Section D-Biological Crystallography 50, 874–877 (1994). doi:10.1107/S0907444994006311.
7. R. Henderson, The potential and limitations of neutrons, electrons and X-rays for atomic resolution microscopy of unstained biological molecules. Q Rev Biophys 28(2), 171–193 (1995).
8. V. Kanike, J. Meesungnoen, J.-P. Jay-Gerin, Acid spike effect in spurs/tracks of the low/high linear energy transfer radiolysis of water: potential implications for radiobiology. RSC Advances 5(54), 43361–43370 (2015). doi:10.1039/c5ra07173a.
9. E.S. Kempner, Effects of high-energy electrons and gamma rays directly on protein molecules. J Pharm Sci 90(10), 1637–1646 (2001). doi:S0022-3549(16)30841-3 [pii].
10. E.S. Kempner, R. Wood, R. Salovey, The temperature dependence of radiation sensitivity of large molecules. Journal of Polymer Science Part B: Polymer Physics 24(10), 2337–2343 (1986). doi:10.1002/polb.1986.090241015.
11. M. Mernea, O. Calborean, O. Grigore, T. Dascalu, D.F. Mihailescu, Validation of protein structural models using THz spectroscopy: a promising approach to solve three-dimensional structures. Optical and Quantum Electronics 46(4), 505–514 (2014). doi:10.1007/s11082-013-9872-0.
12. M.H. Olsson, C.R. Sondergaard, M. Rostkowski, J.H. Jensen, PROPKA3: Consistent Treatment of Internal and Surface Residues in Empirical pKa Predictions. J Chem Theory Comput 7(2), 525–537 (2011). doi:10.1021/ct100578z.
13. R.B.G. Ravelli, S.M. McSweeney, The 'fingerprint' that X-rays can leave on structures. Structure with Folding & Design 8(3), 315–328 (2000). doi:10.1016/S0969-2126(00)00109-X.
14. M. Weik, R.B. Ravelli, G. Kryger, S. McSweeney, M.L. Raves, M. Harel, P. Gros, I. Silman, J. Kroon, J.L. Sussman, Specific chemical and structural damage to proteins produced by synchrotron radiation. Proc Natl Acad Sci U S A 97(2), 623–628 (2000).

Chapter 7
Compressive Sensing Imaging at Sub-THz Frequency in Transmission Mode

Vedat Ali Özkan, Yıldız Menteşe, Taylan Takan, Asaf Behzat Şahin, and Hakan Altan

Abstract Due to lack of widespread array imaging techniques in the THz range, point detector applications coupled with spatial modulation schemes are being investigated using compressive sensing (CS) techniques. CS algorithms coupled with innovative spatial modulation schemes which allow the control of pixels on the image plane from which the light is focused onto single pixel THz detector has been shown to rapidly generate images of objects. Using a CS algorithm, the image of an object can be reconstructed rapidly. Using a multiplied Schottky diode based multiplied millimeter wave source working at 113 GHz, a metal cutout letter F, which served as the target was illuminated in transmission. The image is spatially discretized by laser machined, 10 × 10 pixel metal apertures to demonstrate the technique of spatial modulation coupled with compressive sensing. The image was reconstructed by modulating the source and measuring the transmitted flux through the metal apertures using a Golay cell. Experimental results were compared to reference image to assess reconstruction performance using χ^2 index. It is shown that a satisfactory image is reconstructed below the Nyquist rate which demonstrates that after taking into account the light intensity distribution at the image plane, compressive sensing is an advantageous method to be employed for remote sensing with point detectors.

Keywords Terahertz • Compressive Sensing • Imaging

V.A. Özkan • Y. Menteşe • T. Takan • H. Altan (✉)
Department of Physics, Middle East Technical University, Ankara 06800, Turkey
e-mail: haltan@metu.edu.tr

A.B. Şahin
Department of Electronics and Communication Engineering, Yıldırım Beyazıt University, Ankara 06030, Turkey

M.F. Pereira, O. Shulika (eds.), *THz for CBRN and Explosives Detection and Diagnosis*, NATO Science for Peace and Security Series B: Physics and Biophysics, DOI 10.1007/978-94-024-1093-8_7

7.1 Introduction

There are many applications of Terahertz and millimeter waves such as security screening and quality control [1, 2]. Array imaging systems have still not quite had sufficiency of sensitivity to determine concealed objects with high accuracy. Furthermore, by the consequence of engineering limitations the systems used for these kind of applications are not sufficient for reconstructing images at video rate [3]. In order to achieve video rate imaging, innovations in optical design have been investigated including vibrating mirrors [3, 4] and high sensitivity array based receivers [5].

A point detector coupled with spatial modulation schemes using compressive sensing (CS) analysis routines offers an alternative way to achieve video rate imaging. CS changes the data acquisition and image reconstruction algorithms by granting the system to use less data to obtain the same image rather than changing the optical design [6, 7]. In other words, using CS based algorithms makes it possible to build single pixel cameras [8, 9].

The purpose of this study is to reconstruct the image of an object by using CS below the Nyquist limit. To achieve this purpose, a highly reflective and metal negative target (cut-out letter F) was used. The imaging was realized in transmission mode where optical system was constructed based on making an optimization of image plane location with the help of geometrical ray tracing algorithms. For reconstruction of image l_1 minimization algorithm was used. For evaluation of the reconstructed image the χ^2 image quality assessment (IQA) index was used. The obtained image was processed by using common image processing algorithms.

7.2 Theory

7.2.1 Reconstruction Algorithm

To reconstruct images with spatial light modulators, the following matrix equation needs to be solved

$$y_i = \sum_{j=1}^{M} A_{ij} x_j \quad \textit{for } 1 \leq i \leq N \tag{7.1}$$

where y_i is the measurement taken with i^{th} mask, x vector is the reconstructed image, A_{ij} is the mask's j^{th} pixel's transmission to the receiver, N is the total number of the masks used and M is the total number of pixels. In CS the number of unknowns are greater than the number of equations, $N < M$, so the solution is not unique. Therefore a complimentary method is needed. Several methods are available, such as total variation (TV) minimization [10]. In this study l_l norm minimization method was used [11].

l_1 norm minization method reconstructs the image by the following equation

$$x = \mathrm{arg}min\|x\|_1 \; s.t.|Ax - y|_2 < \varepsilon \tag{7.2}$$

here minimization is in the l_1 metric and the ε is the given amount of error on the measurements.

7.2.2 Image Processing Algorithms

The images obtained by reconstruction were processed for localizing possible areas of interest in the field of view. Even though there was no barrier present in this work, the algorithms were developed for imaging through a barrier.

The first step in processing the reconstructed image was convolving it with a discrete Gaussian kernel. Then the convolved image was gamma corrected for optimal human perception.

7.2.3 Image Quality Assessment

To assess the accuracy of our reconstruction method a quantitative description for the similarity between the obtained images and the ideal image is needed. In this work we used the χ^2 metric for this task. Structural Similarity (SSIM) index is an alternative method for this comparison [12]. Due to the fact that a pixel is 1/100th of the image, a simpler metric which takes only pixel-by-pixel difference into account is preferred [13]. For two images I and R, χ^2 is

$$\chi^2 = \sum_{i,j} \omega_{ij} \left(I_{ij} - R_{ij}\right)^2, \tag{7.3}$$

where, ω_{ij} is the weight of (ij)th pixel. For the purposes of this study, pixels are considered to have equal weight and $\chi^2$is normalized to be unity when considering the similarity between an all-white image, i.e. $I_{ij} = 1$ and an all-black image, i.e. $R_{ij} = 0$ by choosing

$$\omega_{ij} = \frac{1}{M}, \forall i, j.$$

7.3 Experimental Setup

The set-up shown in Fig. 7.1 is used to demonstrate that a highly reflective, metal negative surface can be imaged. The optical system between the target and the masks inverts the target without altering its size. An f#2 Teflon lens with a diameter of 50 mm was used in order to collimate the sub-THz waves originating from the source. The spatial light modulator and the target are separated by two f#2 Teflon lenses, both with 50 mm diameter, to minimize optical sources of error. Finally an f#2 Teflon lens with 50 mm diameter focuses the light on receiver. The image plane position of the optical system was determined using geometrical ray tracing.

A modified Virginia Diode, Inc.(VDI), Schottky diode based, multiplied source with an average power of about 20 mW operating at 113 GHz is used as the sub-THz frequency signal source. Transmitted signals are amplitude modulated via the TTL input by a function generator at 40 Hz and the received signals are detected by a Golay Cell (Tydex TC-1T). The signal intensities were recorded through a 100 MHz digital oscilloscope which was externally gated by the modulation on the source.

The masks and the target were cut from a 0.05 mm thick aluminum sheet using a 40 kHz repetition rate pulsed fiber laser with 0.5 mJ average pulse energy and 25 kW peak power per pulse (Fiberlast NanoMark 20 W). Inverted visible image of the target is given in Fig. 7.2.

The size of the metal target and metal masks were chosen according to the radiation field on the target plane, so that the target can be fully illuminated by the radiation field. Since the diameter of the lenses were chosen to be 5 cm, the size of the mask was limited to 3.2 cm. Considering the 0.2 mm wire frame thickness, each mask was patterned with holes of inner dimensions 3 × 3 mm. Therefore the resolution of the imaging system was determined as 10 × 10. The light intensity distribution at the position of the image plane was evaluated by examining the point spread function distribution across an area of 3.2 cm by 3.2 cm within the size limits of the metal masks. This information was used in reconstructing the image, detailed in Eqs. 7.1 and 7.2.

In total 93 masks were used in measurement. Each mask was designed with the constraint that 50 random holes, out of possible 100, are cut. This constraint enabled to maximize both linear independence of masks and the signal to noise ratio of the measurements.

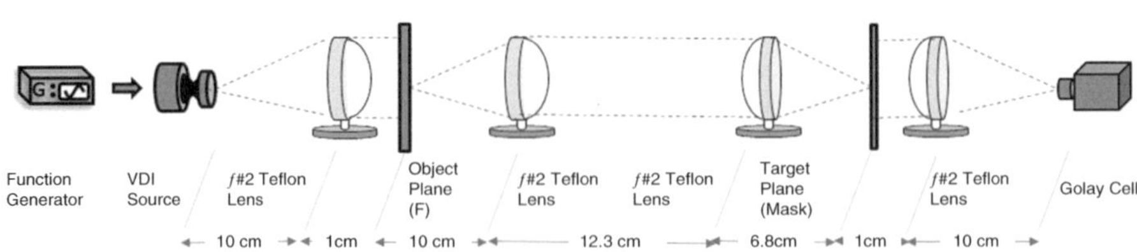

Fig. 7.1 The setup used in the experiment

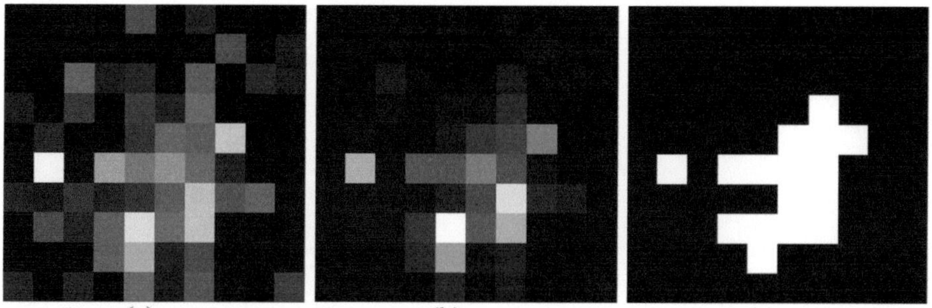

Fig. 7.2 Image reconstructed after 82 measurements (**a**), after processing (**b**), after thresholding (**c**)

During the experiment both the target (negative cut-out F) and the mask being used were placed in polyethylene holders. The masks were changed one by one manually. The received signals were recorded and then fed into Eqs. 7.1 and 7.2 for reconstruction.

7.4 Results

After each measurement, the image was reconstructed as detailed in Sect. 7.2.1. Then each image was compared to the reference image given in Fig. 7.3 using IQA index, detailed in Sect. 7.2.3. Fig. 7.4 gives the evolution of error (χ^2). χ^2 IQA index is useful for comparing sparse pixelated images as the ones obtained here [13].

The evolution of the error though sequential addition of each metal mask is in agreement with CS reconstruction principles. Initially error is decreasing, then stabilizing and then increasing. It should be noted that due to the errors associated with manual positioning of each mask which was within an accuracy of ± 1 mm the error of the model is unpredictable. This in turn means the chosen ε in Eq. 7.2 was incorrect, which negatively influences the image reconstruction. This unpredictable ε error is probably the main cause of the deviations from the general decreasing trend.

After each measurement, the image was reconstructed as detailed in Sect. 7.2.1. Then the reconstructed images were processed as detailed in Sect. 7.2.2. The reconstructed images are analyzed at the 82nd mask which is the minima of the error analysis comparison as shown in Fig. 7.4. After reconstructing the obtained image at this mask position, the image is first processed and then thresholded. These results are given respectively in Fig. 7.2.

Fig. 7.3 Reference image used to assess the image quality index of the reconstructed measurements

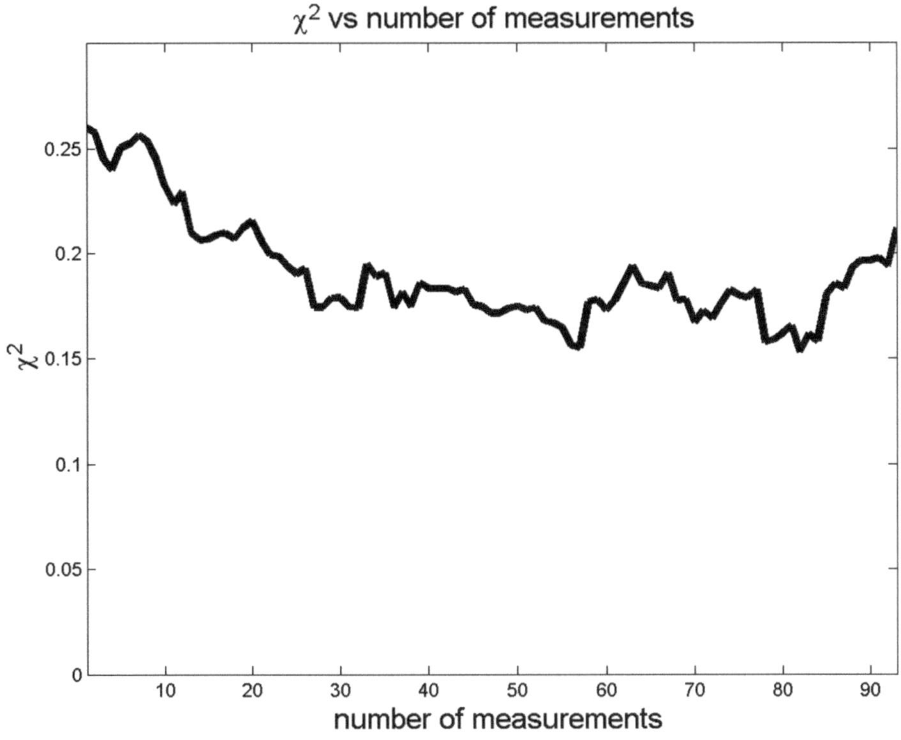

Fig. 7.4 Image quality index (IQA) based on χ^2 error between reference image and reconstructed measurements

7.5 Conclusion

In this paper, CS was used to obtain image of a metal negative target in transmission mode below the Nyquist limit and the obtained image was processed. Although the optical setup was optimized to minimize optical sources of error, manual placement of masks created a mismatch between measurement model, A_{ij} and reality, causing error in reconstruction. Nevertheless the image was successfully obtained below Nyquist limit as shown in Fig. 7.4.

Acknowledgments This project was funded under Turkish Ministry Of cience, Industry and Technology grant #0994STZ2011-2. The author would also like to acknowledge the support of COST MP1204 Action "Tera-MIR Radiation: Materials, Generation, Detection and Applications".

References

1. T. Takan, V. A. Özkan, F. İdikut, İ. O. Yıldırım, A. B. Şahin and H. Altan, Compressive sensing imaging through a drywall barrier at sub-THz and THz frequencies in transmission and reflection modes, Proc. SPIE 9244, Image and Signal Processing for Remote Sensing XX, 92441N (2014)
2. W. L. Chan, J. Deibel, and D. M. Mittleman, "Imaging with terahertz radiation", Reports on Progress in Physics 70(8), p. 1325 (2007)
3. K. Cooper, R. Dengler, N. Llombart, T. Bryllert, G. Chattopadhyay, I. Mehdi, and P. Siegel, "An approach for sub-second imaging of concealed objects using terahertz (thz) radar", Journal of Infrared, Millimeter, and Terahertz Waves, 30(12), pp. 1297–1307 (2009)
4. N. Rothbart, H. Richter, M. Wienold, L. Schrottke, H. Grahn, and H.W. Hubers, "Fast 2-d and 3-d terahertz imaging with a quantum-cascade laser and a scanning mirror", Terahertz Science and Technology, IEEE Transactions on 3(5), pp. 617–624 (2013)
5. D. Rozban, A. Levanon, H. Joseph, A. Akram, A. Abramovich, N. Kopeika, Y. Yitzhaky, A. Belenky, and O. Yadid-Pecht, "Inexpensive thz focal plane array imaging using miniature neon indicator lamps as detectors," Sensors Journal, IEEE, 11(9), pp. 1962–1968 (2011)
6. D. Donoho, "Compressed sensing", Information Theory, IEEE Transactions on, 52(4), pp. 1289–1306 (2006)
7. E. Candes, J. Romberg, and T. Tao, "Robust uncertainty principles: exact signal reconstruction from highly incomplete frequency information", Information Theory, IEEE Transactionsions on, 52(2), pp. 489–509 (2006)
8. M. Duarte, M. Davenport, D. Takhar, J. Laska, T. Sun, K. Kelly, and R. Baraniuk, "Singlepixel imaging via compressive sampling", Signal Processing Magazine, IEEE 25(2), pp. 83–91 (2008)
9. W. L. Chan, M. L. Moravec, R. G. Baraniuk, and D. M. Mittleman, \Terahertz imaging with compressed sensing and phase retrieval," Opt. Lett. 33, pp. 974–976 (2008)
10. L. I. Rudin, S. Osher, and E. Fatemi, "Nonlinear total variation based noise removal algorithms," Phys. D 60, pp. 259–268 (1992)
11. D. L. D. Scott Shaobing Chen and M. A. Saunders, "Atomic decomposition by basis pursuit," SIAM Review, 43(1), pp. 129–159 (2001)
12. Z. Wang, A. Bovik, H. Sheikh, and E. Simoncelli, "Image Quality Assessment: From Error Visibility to Structural Similarity," IEEE Transactions on Image Processing, 13, pp. 600–612 (2004)
13. Y. Yitzhaky, E. Peli, "A method for objective edge detection evaluation and detector parameter selection", in Pattern Analysis and Machine Intelligence, IEEE Transactions on, 25(8), pp.1027–1033 (2003)

Chapter 8
Testing of Sub-THz Properties of Bioliquids Using WGM Resonator with Microfluidic Channel

I.A. Protsenko, A.A. Barannik, A.I. Gubin, N.T. Cherpak, and S.A. Vitusevich

Abstract In the present work, the measurement cell for investigation of small-volume liquids in sub-THz range is proposed. The proposed measurement cell is based on the whispering-gallery mode quartz resonator covered with the plastic layer comprising the microfluidic channel and allows obtaining the complex permittivity of the liquid filling the microfluidic channel. Investigation of the resonator structure was carried out experimentally and using the numerical model of the structure. It is shown that the resonator structure can be used as a measurement cell with additional calibration procedure.

Keywords Complex permittivity • Microfluidic channel • Quartz resonator • Whispering-gallery modes

8.1 Introduction

The permittivity measurements of small-volume liquids is the important area of material testing in the THz frequency range. The obtained permittivity value of solution make it possible to identify the impurity substances in the liquid or estimate the solution concentration change. The whispering-gallery mode (WGM) resonators make possible to obtain the small-volume liquid permittivity with high accuracy, and it was shown earlier in Ka-band [1, 2]. So, the measurement cells based on such resonators may be used for monitoring of bioliquids and chemical

---

I.A. Protsenko (✉) • A.A. Barannik • A.I. Gubin • N.T. Cherpak
Department of Solid-State Radiophysics, O. Ya. Usikov Institute for Radiophysics and Electronics NAS of Ukraine, Kharkiv 61085, Ukraine
e-mail: i.o.protsenko@gmail.com

S.A. Vitusevich
Peter Grünberg Institute (PGI-8), Forschungszentrum Jülich, Jülich 52425, Germany
e-mail: s.vitusevich@fz-juelich.de

M.F. Pereira, O. Shulika (eds.), *THz for CBRN and Explosives Detection and Diagnosis*, NATO Science for Peace and Security Series B: Physics and Biophysics,
DOI 10.1007/978-94-024-1093-8_8

solutions, potentially, for detecting of explosives via small concentrations of volatile substances.

In the present work, the modified measurement technique based on the dielectric resonator with the microfluidic channel (MFC) in plastic layer allowing obtaining the complex permittivity of small-volume liquids in sub-THz range is proposed.

8.2 Experimental and Modeling Details

In the Fig. 8.1 the measurement cell based on the cylindrical quartz resonator covered with the plastic layer comprising the MFC (Fig. 8.1a) and its numerical model prepared in the commercial software COMSOL Multiphysics (Fig. 8.1b) is shown. The radius of the quartz resonator and its height are $R = 4.44$ mm, $h = 1.25$ mm respectively. The height of the plastic layer is 0.61 mm, the diameter of the MFC is $D_{\mathrm{MFC}} \approx 0.2$ mm. The MFC is placed in the middle of the height of plastic layer. The distance between the resonator center and the MFC is above 3 mm. The liquid under test fills the MFC. The liquid filling the microfluidic channel interacts with the resonator electromagnetic field and has a strong influence on the resonator frequency and Q-factor. This effect allows solving the inverse electromagnetic problem by processing the data in COMSOL and obtaining the permittivity of the liquid filling the microfluidic channel. The experimental setup includes also two dielectric (polyethylene) waveguides, precise positioning system and BWO oscillator with PLL, directional coupler and pair of the detectors (to measure incident and transmitted powers). The setup allows obtaining the frequency and Q factor of the structure in frequency range of 118–178 GHz [3].

The diameter of the MFC, the distance between the MCF, the resonator center and the height of the MFC in the plastic layer are difficult for the precise measuring. Therefore, these dimensions should be corrected. For this purpose, the results of the experiment were compared with the numerical ones changing the dimensions mentioned above in the numerical model. This correction was performed for the resonator with the water-filled MFC, therefore the measurement cell in this case is more suitable for the research of different aqueous solutions.

8.2.1 Choosing the Operating Mode

The plastic layer does not significantly change the spectrum of the quartz disk. Therefore, at the beginning the properties of the bare quartz disk were studied in the sub-THz range. The spectrum of the quartz disk was obtained experimentally, and then modes were identified and the eigenfrequencies were calculated using the numerical model. Resonance lines and field pattern of some WGM in the range of

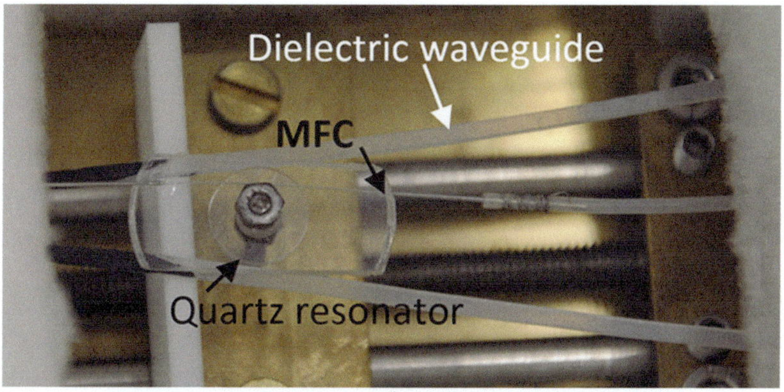

a)

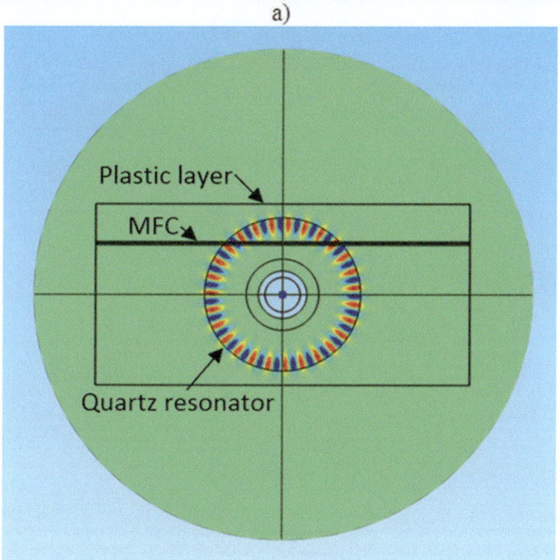

b)

Fig. 8.1 The measurement cell based on the quartz resonator covered with the plastic layer comprising the MFC (**a**) and its numerical model prepared in the COMSOL Multiphysics (**b**)

165–175 GHz are shown in the Fig. 8.2. The resonator with the mode HE $_{27,\ 1,\ 1}$ (Fig. 8.2) has the highest Q-factor ($Q \approx 32{,}200$), and this mode was chosen as an operating one. The field pattern of the mode HE $_{27,\ 1,\ 1}$ in the resonator with the MFC is shown in the Fig. 8.1b.

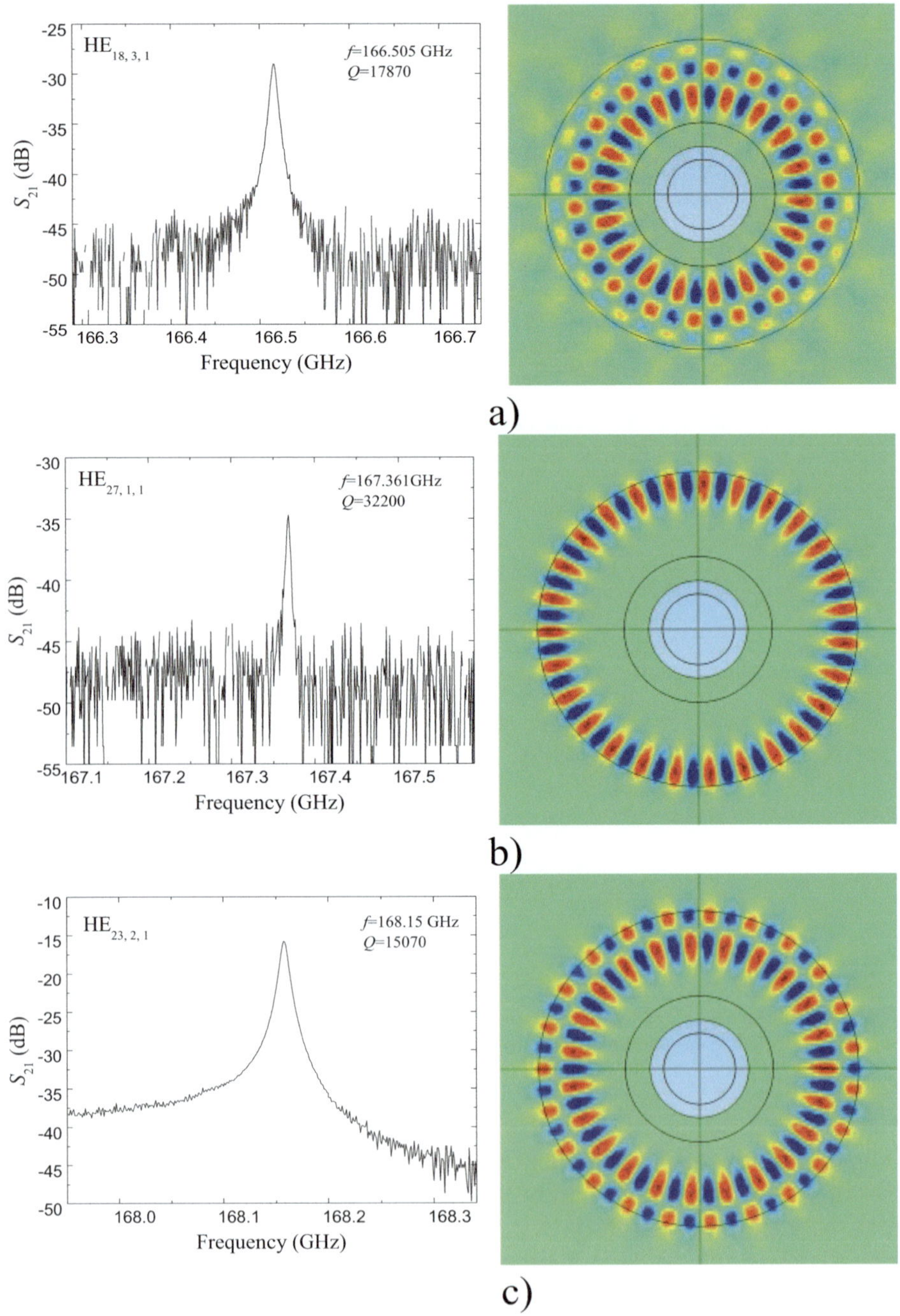

Fig. 8.2 The resonance lines and field pattern for different WGM: HE $_{18,\ 3,\ 1}$ (**a**), HE $_{27,\ 1,\ 1}$ (**b**), HE $_{23,\ 2,\ 1}$ (**c**)

Table 8.1 Numerical vs experimental frequencies and Q-factors

	Frequency, GHz		*Q*-factor	
	Experiment	Numerical	Experiment	Numerical
Bare quartz disk	167.36	167.36	32,200	34,000
Quartz resonator with the air-filled MFC	166.466	166.566	19,120	19,066

8.2.2 Study of the Resonator with the MFC Filled with Liquids

Frequency and *Q*-factor for both bare quartz disk and quartz disk covered with the plastic Zeonex layer comprising the air-filled MFC with the HE $_{27,\ 1,\ 1}$ mode are shown in the Table 8.1. The air-filled MFC, unlike from the MFC filled with a high-losses liquid, does not cause the significant change of the resonator characteristics. However, presence of the plastic layer causes the decreasing of the resonant frequency (about 900 MHz). Plastic layer also involves additional losses, therefore the resonator *Q*-factor increases, but its value is suitable for the measurements anyway.

As it was mentioned above, the MFC diameter and its place in the plastic layer had been corrected using the results of experiment and the numerical investigations for the case of water-filled MFC. Then the numerical model with the corrected dimensions with the MCF filled with the ethanol was considered to make sure that the numerical model is identical to the resonator structure used in the experiment. The permittivity of water and ethanol used in the numerical model has been obtained in [4, 5]. The frequency shift (between the frequencies of the resonator with the air-filled MFC and the resonator with the MFC filled with liquid) and the difference in inverse *Q*-factor of the resonator with the MFC filled with both water and ethanol and the resonator with the air-filled MFC are shown in the Table 8.2.

As one can see in Table 8.2, the results of the experiment and calculation are in a good agreement. Hence, the resonator structure can be used as a measurement cell for obtaining permittivity of small-volume liquids filling the MFC. To this purpose the experimental data (frequency shift and inverse *Q*-factor) should be proceeded with the COMSOL model. The specific calibration procedure can be used for this purpose [1].

8.3 Conclusion

The resonator structure based on the WGM quartz resonator covered with the plastic layer with the MFC is described and studied. The spectrum of the resonator was obtained in sub-THz range and the operating high-*Q*-factor mode was chosen. The influence of the plastic layer and the MFC filled with liquids on the resonator characteristics is shown. The measurement cell was studied both numerically using

Table 8.2 Numerical vs experimental frequency shifts and inverse Q-factors

	Frequency shift, $f_{air} - f_{liquid}$, MHz		Inverse Q-factors $Q^{-1}{}_{liquid} - Q^{-1}{}_{air}$	
	Experiment	Numerical	Experiment	Numerical
Water	7.135	7.150	6.091×10^{-5}	6.115×10^{-5}
Ethanol	3.45	3.20	8.674×10^{-6}	9.251×10^{-6}

commercial software COMSOL and experimentally. The results are in good agreement for water- and ethanol-filled MFC. It is shown that the structure can be used as a measurement cell to obtain the bioliquids permittivity.

Acknowledgments The work was supported in part by a DFG project number GZ: VI 456/3-1.

References

1. A. I. Gubin, A. A. Barannik, N. T. Cherpak, I. A. Protsenko, S. Pud, A. Offenhaeusser, and S. A. Vitusevich, Whispering-gallery-mode resonator technique with microfluidic channel for permittivity measurements of liquids. IEEE Trans. MTT 63, 2003–2009 (2015). doi: 10.1109/TMTT.2015.2438392.
2. J. Krupka, K. Derzakowski, B. Riddle, J. Baker-Jarvis, A dielectric resonator for measurements of complex permittivity of low loss dielectric materials as a function of temperature. Meas. Sci. Technol. 9, 1751–1756 (1998). doi: 10.1088/0957-0233/9/10/015.
3. A. I. Gubin, A. A. Barannik, N. T. Cherpak, I. A. Protsenko, S. Pud, A. Offenhaeusser, and S. A. Vitusevich, WGM resonators with microfluidic channel for sub-mm wave characterization of biological liquids. GeMiC 2016, March 2016, Bochum, Germany, Symposium proceeding, 15–18 (2016).
4. W. J. Ellison, Permittivity of pure water, at standard atmospheric pressure, over the frequency range 0–25 THz and the temperature range 0–100 °C. J. Phys. Chem. Ref. Data 36, 1–18 (2007). doi:10.1063/1.2360986.
5. J. Barthel and R. Buchner, High frequency permittivity and its use in the investigation of solution properties. Pure Appl. Chem. 63, 1473–1482 (1991). doi: 10.1351/pac199163101473.

Chapter 9
Optical and THz Evaluation of Components for Gas Sensing Spectroscopy in Hazardous Environments

Dan Sporea, Laura Mihai, Adelina Sporea, Daniel Ighigeanu, and Daniel Neguţ

Abstract In this paper, we report for the first time a comparative investigation on gamma and electron beam irradiation effects on some mid-IR materials (ZnSe, CaF_2, BaF_2, and sapphire) used as components in trace gas detection systems, in the 15–100 THz spectral range. Optical transmittance/reflectance measurements were performed from UV to far IR (0.2–23 μm) on both non-irradiated and irradiated samples. These investigations were complemented by THz spectroscopy and imaging tests run under the same conditions.

Keywords THz spectroscopy • THz imaging • Mid-IR spectroscopy • Radiation effects • Gamma-ray • Electron beam

9.1 Introduction

At European level there is an increase concern on existing risks and an awareness to "prevent chemical, biological, radiological, nuclear (CBRN) and explosives incidents and protect citizens, institutions and infrastructure against such incidents" [1]. The existing situation conducted to the preparation of specific action plans to address these issues [2, 3]. These documents highlight the development of detection technologies as one of the major direction to act, implying vehicle, cargo, personnel/passenger screening, according to specific criteria and targeting minimum detecting amounts, to be supported by research, testing, validation and practical trials, targeting the setting of appropriate detection standards and the identification of good practice in the field [4].

D. Sporea (✉) • L. Mihai • A. Sporea • D. Ighigeanu
National Institute for Laser, Plasma and Radiation Physics, Magurele, Romania
e-mail: dan.sporea@inflpr.ro

D. Neguţ
"Horia Hulubei" National Institute of Physics and Nuclear Engineering, Magurele, Romania

M.F. Pereira, O. Shulika (eds.), *THz for CBRN and Explosives Detection and Diagnosis*, NATO Science for Peace and Security Series B: Physics and Biophysics,
DOI 10.1007/978-94-024-1093-8_9

In the last years, terahertz (THz) and mid infrared (mid-IR) spectroscopy proved to be a promising technology for the detection of CBRN and explosives, based on their spectral signature in these spectral intervals [5–7]. The proposed detection schemes included: THz Time Domain Spectroscopy [8], THz imaging [9], the use of quantum cascade lasers (QCL) in active/passive, transmission or reflection spectroscopy [10], photoacoustic detection [11], broadband/hyperspectral standoff imaging [12–14].

An additional challenge comes out when such detection has to be performed under hazardous conditions (e.g. in the presence of ionizing radiation).

In this paper, we report for the first time a comparative investigation on gamma and electron beam irradiation effects on some mid-IR materials for optical components (ZnSe, CaF_2, BaF_2, and sapphire – Al_2O_3) used as components in trace gas detection systems, in the 15–100 THz spectral range. Previous reports on the irradiation of such optical materials refer to gamma-ray, heavy ions or neutron produced effects [15–17]. These reported results addressed mostly the radiation induced changes in the UV to near-IR (up to 1 μm) spectral range.

9.2 Materials and Experiment

Four IR optical materials (CaF_2, BaF_2, ZnSe, and sapphire) were tested under comparable gamma-ray and electron beam exposure. All the evaluated materials and devices are Commercial Off-The-Shelf (COTS) products. For the case of gamma irradiation the optical windows had diameters between 10 and 15 mm and thickness of 1–3 mm, and in the case of electron beam radiation the windows had diameters of 25 mm and thickness of about 2 mm.

Gamma irradiation was performed at the IRASM ^{60}Co irradiator. The dose rate was 5.7 kGy/h, and the total doses were 0.1 kGy, 1 kGy, 10 kGy and 20 kGy. As they were subjected to gamma radiation the samples were kept into plastic bags to avoid any contamination which might affect their optical properties. Electron beam irradiation was done at the 5.5 MeV linear accelerator, pulse duration of 4 μs, operating current 8 μA, frequency of 50 Hz. The dose rate was 4 kGy/min, and the total doses were 7 kGy, 2.2 kGy, 5.2 kGy and 10 kGy. During the electron beam irradiation the optical components were packed into a thin Al foil for heat dissipation and to protect them from the dust.

Prior to any irradiation and following each irradiation step the spectral transmittance, the spectral diffused and specular reflectance were measured over spectral ranges specific to each material: CaF_2 (0.21–12 μm), BaF_2 (0.2–16 μm), ZnSe (0.46–23 μm), Al_2O_3 (0.2–6.9 μm). These spectral characterizations were performed with the Gooch and Housego OL750 (Fig. 9.1) with the appropriate diffracting gratings, a tungsten lamp and an IR glower lamp covering the spectral range 0.2–21 μm, an AC lock-in amplification system coupled to reference detectors: Si (0.2–1.1 μm), Ge (0.8–1.8 μm), PbS (1.8–3.2 μm), InSb (3–5.5 μm), HgCdTe (5.5–21 μm).

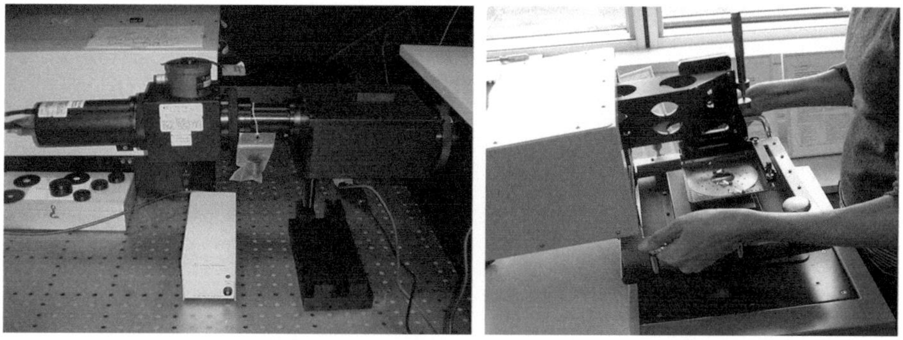

Fig. 9.1 The setup for optical spectral tests run on irradiated mid-IR windows

The optical spectral measurements were complemented by THz spectral transmission investigations and THz imaging using the TeraView TPS Spectra 3000 THz spectrometer with the Reflectance Imaging Module. The THz studies referred to the irradiation induced changes on spectral transmittance, refractive index and spectral absorption coefficient for the four mid-IR materials.

9.3 Results

Figure 9.2 illustrates as an example the modifications of the four optical materials spectral transmittance, as they were irradiated by gamma-ray and electron beam.

Figure 9.3 provides information on the refractive index change of BaF_2 and ZnSe in the THz spectral range (0.2–3 THz) as they respond to electron beam and gamma irradiation.

The electron beam and gamma-ray irradiation induced effects on the absorption coefficient of Al_2O_3 over the 0.2–3 THz interval are reproduced in Fig. 9.4a, as compared to the same characteristics for the non-irradiated sample. In Fig. 9.4b are exemplified the modifications produced by the electron beam and gamma-ray exposure on the spectral transmittance of the CaF_2 window, in the THz spectral range, for different total doses.

Thz imaging was used to monitor irradiation induced defects in the studied mid-IR windows (Fig. 9.5). The frequency domain analysis mode was used for this purpose.

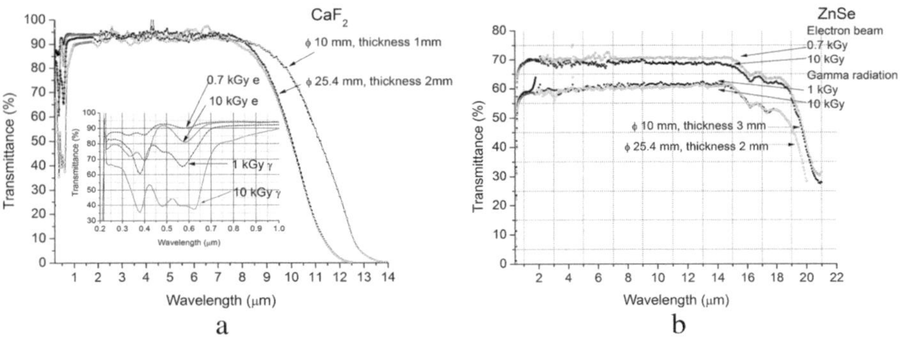

Fig. 9.2 The spectral transmittance of (**a**) CaF_2 and (**b**) ZnSe samples after their irradiation with gamma-ray and electron beam

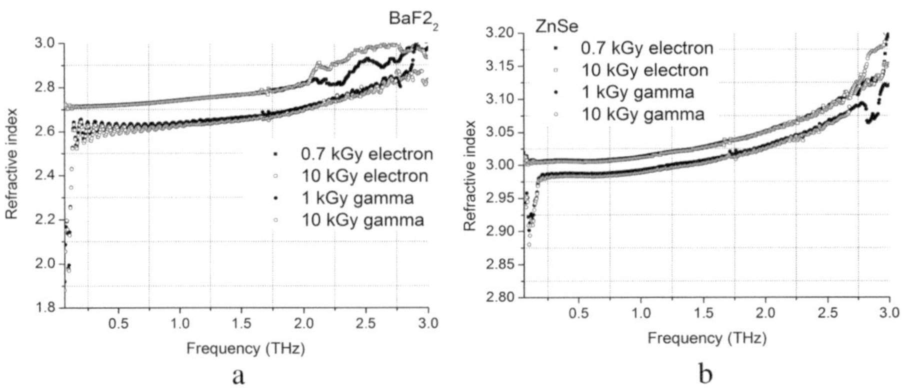

Fig. 9.3 The changes in refractive index after the irradiation of (**a**) BaF_2 and (**b**) ZnSe samples by electron beam and gamma radiation

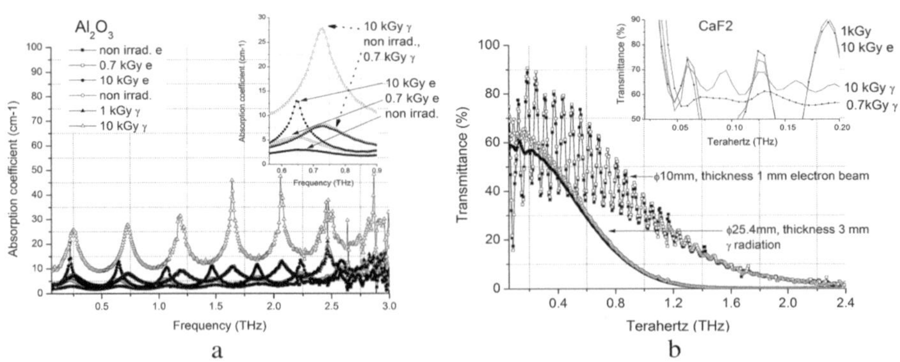

Fig. 9.4 The changes in (**a**) absorption coefficient on sapphire sample and (**b**) transmittance properties of CaF_2 after two different types of irradiations

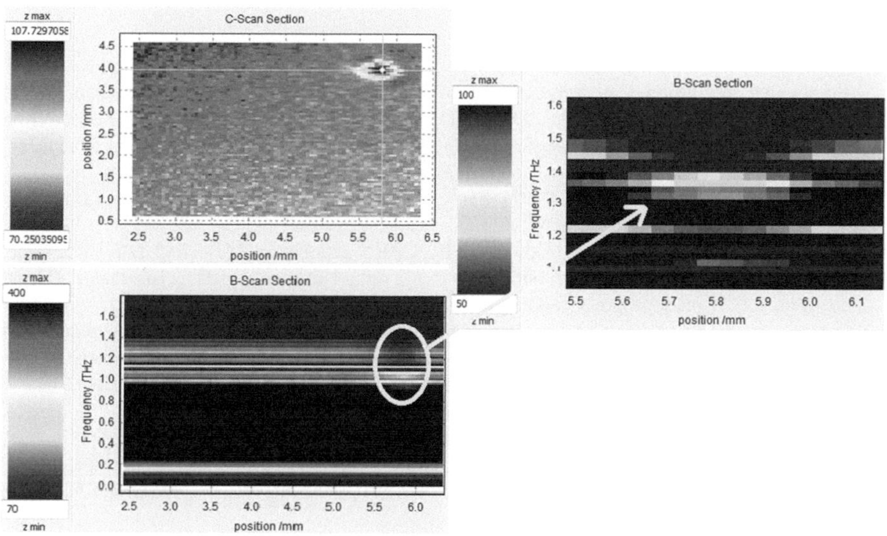

Fig. 9.5 Defect noticed on XY mapping (*C-Scan*) and on reflectance variation with frequency (*B-Scan*) following electron beam irradiation for the CaF_2 sample after a cumulative dose of 10 kGy. The measurements were done with RIM module of TPS3000 spectrometer at a scanning step of 50 microns over a 4×4 mm surface, central to the sample

9.4 Conclusion

In the present investigation we tested four mid-IR window materials for the radiation produced changes in the optical and THz spectral ranges, as these materials were subjected to gamma-ray and electron beam irradiation. Total cumulated dose for gamma-ray exposure was 20 kGy (at a dose rate of 5.7 kGy/h), while the total dose for electron beam irradiation was 10 kGy (dose rate 4 kGy/min).

Modifications of the optical transmittance in UV were noticed for both types of irradiation, the most significant change being found under gamma irradiation in the case of CF_2 (Fig. 9.2a). ZnSe proved to be more radiation resistant over the studied spectral optical range, slight decrease of the optical transmittance was observed in the 5–7 μm and 13–16.5 μm intervals, under electron beam irradiation (Fig. 9.2b). As it concerns the THz refractive index of the BaF_2 and ZnSe windows small changes are present after electron beam exposure (Fig. 9.3). In the case of sapphire the absorption coefficient presents some peaks more evident following gamma irradiation (Fig.9.4a). For CaF_2 the spectral transmittance is more affected after electron beam exposure (Fig. 9.4b). XY mapping of the irradiated samples offers for the first time useful information on the localization of radiation induced defects (Fig. 9.5).

Acknowledgments The authors acknowledge the financial support of Romanian Space Agency (ROSA), ECSA project, contract 67/2013. Some of the equipments used in this research were purchased in the frame of the project "Center for Advanced Lasers Technologies (CETAL)", contract 8PM/2010, financed by UEFISCDI.

References

1. European Commission, Communication from the Commission to the European Parliament, the Council, the European Economic and Social Committee and the Committee of the Regions on a new EU approach to the detection and mitigation of CBRN-E risks. (Brussels, 5.5.2014, COM (2014) 247 final), http://ec.europa.eu/dgs/home-affairs/what-we-do/policies/crisis-and-terrorism/explosives/docs/20140505_detection_and_mitigation_of_cbrn-e_risks_at_eu_level_en.pdf. Accessed 05 March 2016.
2. Progress report on the implementation of the EU CBRN action plan, May 2012. http://ec.europa.eu/dgs/home-affairs/what-we-do/policies/crisis-and-terrorism/securing-dangerousmaterial/docs/eu_cbrn_action_plan_progress_report_en.pdf. Accessed 05 March 2016.
3. Progress report on the implementation of the EU action plan on enhancing the security of explosives. http://ec.europa.eu/dgs/home-affairs/what-we-do/policies/crisis-andterrorism/explosives/docs/progress_report_on_explosives _security_2012_en.pdf. Accessed 05 March 2016.
4. Council of the European Union, Council conclusions on strengthening chemical, biological, radiological and nuclear (CBRN) security in the European Union - an EU CBRN Action Plan. (Brussels, 12 November 2009, 15505/1/09, Rev 1). http://register.consilium.europa.eu/doc/srv?l=EN&f=ST%2015505% 202009%20REV%201%20COR%201. Accessed 05 March 2016.
5. R.M. Woodward, in *Optically Based Biological and Chemical Sensing for Defence*, ed. by J.C. Carrano, A. Zukauskas, Proc. SPIE 5617 (SPIE, 2004), p. 341. doi:10.1117/12.569427.
6. J. Hildenbrand, J. Herbst, J. Wöllenstein, A. Lambrecht, in *Quantum Sensing and Nanophotonic Devices VI*, ed. by M. Razeghi, R. Sudharsanan, G.J. Brown, Proc. SPIE 7222 (SPIE, 2009), p. 72220B. doi:10.1117/12.808976.
7. F.K. Tittel et al., in *Terahertz and Mid Infrared Radiation: Detection of Explosives and CBRN (Using Terahertz)*, ed. by M. Pereira, O. Shulika (Springer, Berlin, Heidelberg, New York, 2014), p. 153.
8. T. Trzcinski, N. Palka, M. Szustakowski, Polska Akademia Nauk. Bulletin of the Polish Academy of Sciences **59**(4), 445 (2011). doi: 10.2478/v10175-011-0056-4.
9. V. L. Vaks, J. Physics: Conference Series **486,** 012002 (2014) doi:10.1088/1742-6596/486/1/012002.
10. E.R. Deutsch et al., in *Advanced Environmental, Chemical, and Biological Sensing Technologies XI*, ed. by T. Vo-Dinh; R.A. Lieberman; G.G. Gauglitz, Proc. SPIE 9106 (SPIE, 2014), p. 91060A. doi: 10.1117/12.2058544.
11. C.W. Van Neste, L.R. Senesac, T. Thundat, Appl. Phys. Lett. **92**, 234102 (2008). doi: 10.1063/1.2945288.
12. M.E. Morales-Rodríguez et al., in *Micro- and Nanotechnology Sensors, Systems, and Applications III*, Proc. SPIE 8031 (SPIE 2011), p. 80312D. doi: 10.1117/12.885204.
13. A. C. Padilla-Jiménez, W. Ortiz-Rivera, C. Rios-Velazquez, I. Vazquez-Ayala, S. P. Hernández-Rivera, Opt. Eng. **53**(6), 061611 (2014). doi: 10.1117/1.OE.53.6.061611.
14. P. Kotidis, E.R. Deutsch, A. Goyal, in *Micro- and Nanotechnology Sensors, Systems, and Applications VII*, ed. by T. George, A.K. Dutta, M. Saif Islam, Proc. SPIE 9467 (SPIE, 2015), p. 94672S. doi: 10.1117/12.2178169.
15. C.L.Woody, J.A. Kierstead, P.W. Levy, S. Stoll, IEEE Trans. Nucl. Sci. **NS-39**, 515 (1992)
16. A.S. El-Said et al., Nucl. Instrum. Meth. B. **218**, 492 (2004). doi: 10.1016/j.nimb.2003.12.057.
17. M. Izerrouken, A, Meftah, M. Nekka, J. Lumin. **127**, 696 (2007). doi: 10.1016/j.jlumin.2007.04.005.

Chapter 10
Analysis of Active Pharmaceutical Ingredients by Terahertz Spectroscopy

Y. Samet Aytekin, Mustafa Köktürk, and Okan Esenturk

Abstract Spectra of solid state active pharmaceutical ingredients obtained by Terahertz (THz) Time Domain Spectrometer in terahertz region of electromagnetic spectrum are presented. Each sample showed unique spectral features in 10.0–90.0 cm^{-1} indicating the potential use of THz spectroscopy for identification and quantification of drugs. Spectroscopic information of drugs would be needed in near future when THz technology further develops for on-site analysis and characterisation.

Keywords Terahertz Time Domain Spectroscopy • THz-TDS • Active Pharmaceutical Ingredients • API

10.1 Introduction

THz region of the electromagnetic spectrum lies in spectral range of 0.1–10.0 THz, corresponding to a range between infrared and microwave regions. This region has drawn attention as it provides important spectroscopic information about materials based on their intermolecular interactions and low frequency motions. This fingerprint information enables identification and quantification of materials. Furthermore, THz waves can penetrate through materials non-invasively. Many non-polar materials like cotton, paper, and plastics are transparent to THz radiation making the technique accessible to samples behind these materials. Application of THz spectroscopy has been benefitted in many areas like security and explosives [1, 2], biology [5], material characterisation [6, 7], and imaging [8, 9]. Recently, it is also gaining an increased attention in characterisation of illicit drugs [3, 4].

In pharmaceutical industry, drugs are meticulously analysed for their bioavailability, stability and solubility. Since most of the active pharmaceutical ingredients

Y.S. Aytekin (✉) • O. Esenturk
Department of Chemistry, Middle East Technical University, Ankara 06100, Turkey
e-mail: ysametu@gmail.com

M. Köktürk
Nobel İlaç San. ve Tic. A.Ş., Düzce 81100, Turkey

M.F. Pereira, O. Shulika (eds.), *THz for CBRN and Explosives Detection and Diagnosis*, NATO Science for Peace and Security Series B: Physics and Biophysics,
DOI 10.1007/978-94-024-1093-8_10

(API) have polymorphic forms, each drug has to be checked and analysed for the correct form. For characterisation and control various methods are utilised like x-ray diffractometry, differential scanning calorimetry, IR spectroscopy, Raman spectroscopy and solid state NMR by pharmaceutical industry [10–13]. In addition to these a new and promising technique called THz spectroscopy also appears to be a good candidate for drug analysis. Recent studies have shown that THz spectroscopy is eligible for distinguishing drugs [14, 15], measuring tablet coating thickness [16], studying porosity of tablet [17].

In this study THz spectra of APIs collected by a home-made Terahertz Time Domain Spectrometer are shown in the region of 10.0–90.0 cm^{-1}. Each API samples showed distinct spectral features in the studied region. The results would form a contribution to newly developing API database that is slowly being formed in THz region.

10.2 Experimental

The APIs of commonly used drugs, Benzocaine, Irbesartan, Paracetamol and Fluoxetine HCl, were obtained from Nobel R&D Center and has been used without any further purification. The samples were prepared in pellet forms having 13 mm diameters. Pure samples were mixed with high density polyethylene (HDPE, Micro Powders Inc.) with a particle size of 10 μm. Sample pellets were prepared as API-PE mixture (5% w/w) under 1 ton of pressure for 20 s.

A homebuilt THz-TDS spectrometer and a commercial FTIR spectrometer was used to cover a range from 0.3 to 3.0 THz (9.9–99.9 cm^{-1}). The THz spectrometer light source is a mode-locked Ti:sapphire oscillator laser (Coherent Mantis-5) with a pulse duration of 80 fs, central wavelength of 800 nm, and a repetition rate of 80 MHz. The laser beam was split into two; THz generation arm and detection arm. In the generation arm, the pump beam is focused onto a photoconductive antenna (BATOP, iPCA-21-05-1000-800-h) to populate GaAs conduction band. The excited electrons are accelerated by an applied bias voltage of 15 V at 8 kHz resulting in THz radiation. The THz beam is guided with help of parabolic mirrors in 8F configuration, in which the light is collimated after the generation, focused onto a sample, collected and collimated with another off-axis mirror, and then focused onto a 500 μm <110> ZnTe crystal (MTI corp.) for THz radiation detection. The 800 nm beam in the detection arm is used for THz mapping on the ZnTe detection crystal. By changing the gating time of the THz and the visible pulses arrival on the ZnTe crystal with a delay stage THz beam profile is recorded with help of quarter waveplate, Wollaston prism, balance photodiode, and lock in amplifier on a computer. (Fig. 10.1).

The THz beam pathway is purged with dry air to reduce water absorptions, which is especially attenuating high frequencies of THz radiation. The humidity was kept well below 1% for all measurements. Obtained signals from both sample and reference were processed with 'TDS Spectrum Viewer [18] to extract optical

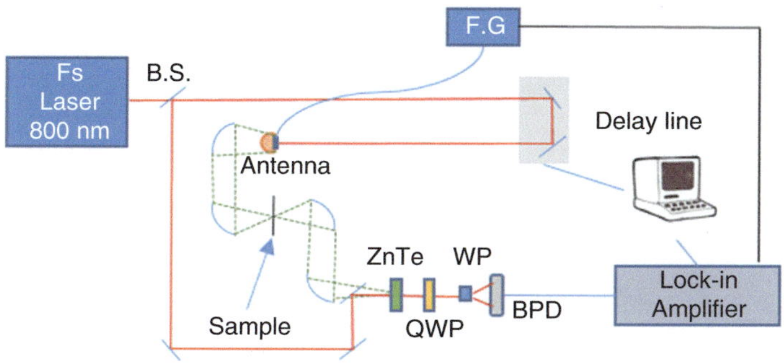

Fig. 10.1 THz time domain spectrometer setup

properties of our samples. All FTIR measurements were carried out by Thermo Nicolet 6700 FTIR spectrometer with an effective working range of 50–700 cm^{-1}. The spectra are collected in the same dry air purged environment.

10.3 Results and Discussion

Figure 10.2 presents the spectra of four different APIs in PE (5%w/w) measured by both THz-TDS and FT-THz systems. Each of the samples has more than one distinct spectral feature that can be used for identification, discrimination, and quantification of these API even in a mixture of samples. The observed peaks are relatively broad as expected in this THz range. Here dispersion effects may also play role in addition to the common band broadening effects.

Three noticeable features are observed at 38 cm^{-1}, 71 cm^{-1} and 92 cm^{-1} for losartan potassium even at 5% level. The first peak is relatively low in intensity and broader compared to the others. FT-THz showed a spectra having peaks located at 71 and 92 cm^{-1} which coincides well with TDS result confirming the observed features with THz-TDS. There are three absorption bands of paracetamol centred at ca. 49, 72, and 95 cm^{-1}. The intensities of the bands increase as the wavenumber increases, and all are resolved. Similarly, the last two peaks (72 and 95 cm^{-1}) are only the ones observed with FT-THz due to the low frequency limit of the FT-THz system. The band observed at ca. 49 cm^{-1} is significantly weaker compared to the others. The characteristic bands of irbesartan are located at 26 and 51 cm^{-1}. Both are relatively strong but quite broad in nature but they could still be discriminated from each other and can be used for identification purposes. All the features are below the common frequency range (50–100 cm^{-1}) thus FT-THz spectrum could not be collected and is not shown in the Fig. 10.2c. Relatively strong but broad absorption bands of benzocaine are observed at 71 and 89 cm^{-1}. The observed bands are well resolved with fairly good intensities that could be utilised for quantitative analysis. These bands are also confirmed in the spectrum obtained by

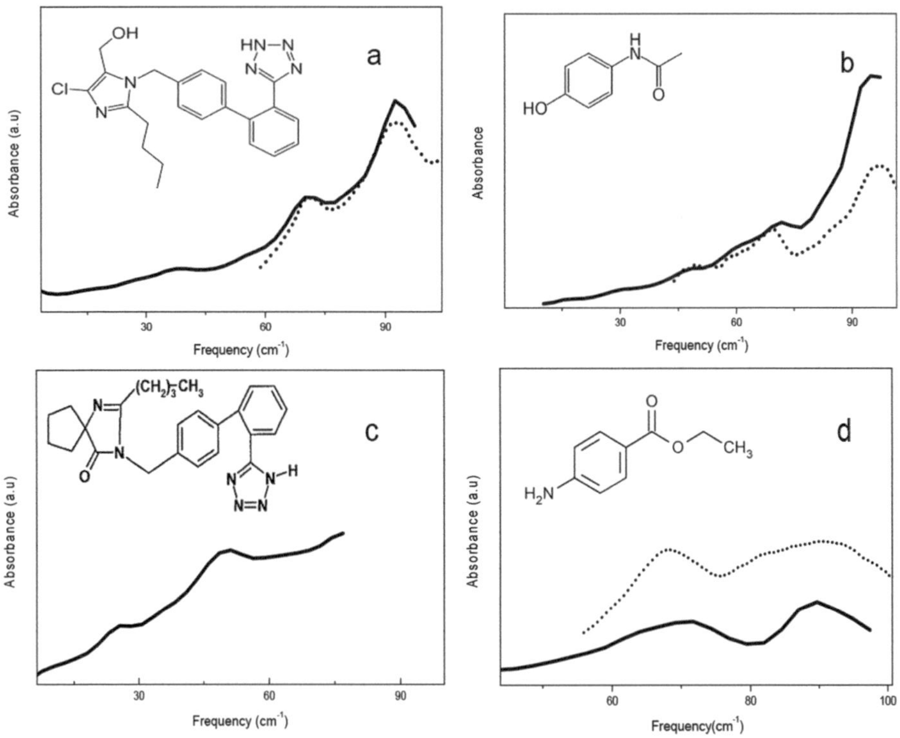

Fig. 10.2 THz-TDS (*black line*) and FT-THz (*short dotted line*) spectra of (**a**) Losartan potassium (**b**) Paracetamol (**c**) Irbesartan (**d**) Benzocaine

FT-THz system. All of the API samples have distinct spectral features at measured frequency range of 10–100 cm^{-1} indicating the potential of the THz based technique for on site, in line or in situ analysis of drugs.

10.4 Conclusion

Optical properties of APIs of four commonly used drugs were presented in THz region. Each API possesses unique THz spectrum that could be used for characterisation, identification and quantification. The spectra are intended to be a contribution to spectral database of API in THz region. In near future with the advancements in this field, it would be possible to analyse the drugs in their packages without using destructive analysis techniques thanks to the penetrating and fingerprinting features of THz radiation.

Acknowledgments This study was financially supported by the Scientific and Technological Research Council of Turkey (TUBITAK) with the project number of 111T393.

References

1. Melinger, J. S.; Laman, N.; Grischkowsky, D. The Underlying Terahertz Vibrational Spectrum of Explosives Solids. *Appl. Phys. Lett.* **2008**, *93* (1), 2006–2009.
2. Fallis, A. . Security Applications of Terahertz Technology. *J. Chem. Inf. Model.* **2013**, *53* (9), 1689–1699.
3. Hakey, P. M.; Allis, D. G.; Ouellette, W.; Korter, T. M. Cryogenic Terahertz Spectrum of (+)-Methamphetamine Hydrochloride and Assignment Using Solid-State Density Functional Theory. *J. Phys. Chem. A* **2009**, *113* (17), 5119–5127.
4. Lu, M.; Shen, J.; Li, N.; Zhang, Y.; Zhang, C.; Liang, L.; Xu, X. Detection and Identification of Illicit Drugs Using Terahertz Imaging. *J. Appl. Phys.* **2006**, *100* (10), 1–6.
5. Ashworth, P. C.; Pickwell-MacPherson, E.; Provenzano, E.; Pinder, S. E.; Purushotham, A. D.; Pepper, M.; Wallace, V. P. Terahertz Pulsed Spectroscopy of Freshly Excised Human Breast Cancer. *Opt. Express* **2009**, *17* (15), 12444–12454.
6. Docherty, C. J.; Johnston, M. B. Terahertz Properties of Graphene. *J. Infrared, Millimeter, Terahertz Waves* **2012**, *33* (8), 797–815.
7. Naftaly, M.; Miles, R. E. Terahertz Time-Domain Spectroscopy for Material Characterization. *Proc. IEEE* **2007**, *95* (8), 1658–1665.
8. Fitzgerald, A. J.; Cole, B. E.; Taday, P. F. Nondestructive Analysis of Tablet Coating Thicknesses Using Terahertz Pulsed Imaging. *J. Pharm. Sci.* **2005**, *94* (1), 177–183.
9. Shen, Y. C.; Taday, P. F.; Newnham, D. A.; Pepper, M. Chemical Mapping Using Reflection Terahertz Pulsed Imaging. *Semicond. Sci. Technol.* **2005**, *20* (7), S254–S257.
10. Riekes, M. K.; Pereira, R. N.; Rauber, G. S.; Cuffini, S. L.; Campos, C. E. M. de; Silva, M. A. S.; Stulzer, H. K. Polymorphism in Nimodipine Raw Materials: Development and Validation of a Quantitative Method through Differential Scanning Calorimetry. *J. Pharm. Biomed. Anal.* **2012**, *70*, 188–193.
11. Yang, C.; Mitra, A. K. Low-Level Determination of Polymorph Composition in Physical Mixtures by near-Infrared Reflectance Spectroscopy. *J. Pharm. Sci.* **2001**, *90* (3), 360–370.
12. Seliger, J.; Žagar, V.; Apih, T.; Gregorovič, A.; Latosińska, M.; Olejniczak, G. A.; Latosińska, J. N. Polymorphism and Disorder in Natural Active Ingredients. Low and High-Temperature Phases of Anhydrous Caffeine: Spectroscopic (1H–14N NMR–NQR/14N NQR) and Solid-State Computational Modelling (DFT/QTAIM/RDS) Study. *Eur. J. Pharm. Sci.* **2016**, *85*, 18–30.
13. Docoslis, A.; Huszarik, K. L.; Papageorgiou, G. Z.; Bikiaris, D.; Stergiou, S.; Georgarakis, E. Characterization of the Distribution, Polymorphism, and Stability of Nimodipine in Its Solid Dispersions in Polyethylene Glycol by Micro-Raman Spectroscopy and Powder X-Ray Diffraction. *AAPS J.* **2007**, *9* (3), E361–E370.
14. Strachan, C. J.; Rades, T.; Newnham, D. a.; Gordon, K. C.; Pepper, M.; Taday, P. F. Using Terahertz Pulsed Spectroscopy to Study Crystallinity of Pharmaceutical Materials. *Chem. Phys. Lett.* **2004**, *390* (1–3), 20–24.
15. Ge, M.; Liu, G.; Ma, S.; Wang, W. Polymorphic Forms of Furosemide Characterized by THz Time Domain Spectroscopy. *Bull. Korean Chem. Soc.* **2009**, *30* (10), 2265–2268.
16. Lin, H.; May, R. K.; Evans, M. J.; Zhong, S.; Gladden, L. F.; Shen, Y.; Zeitler, J. A. Impact of Processing Conditions on Inter-Tablet Coating Thickness Variations Measured by Terahertz In-Line Sensing. *J. Pharm. Sci.* **2015**, *104* (8), 2513–2522.
17. Peiponen, K. E.; Silfsten, P.; Pajander, J.; Ketolainen, J. Broadening of a THz Pulse as a Measure of the Porosity of Pharmaceutical Tablets. *Int. J. Pharm.* **2013**, *447* (1–2), 7–11.
18. http://www.riken.jp/lab-www/THz-img/hoshina/software_e.html; TDS spectrum viewer.

Chapter 11
Fabrication of Bi2212 Single Crystal Bolometer for Detection of Terahertz Waves

T. Semerci, Y. Demirhan, N. Miyakawa, H.B. Wang, and L. Ozyuzer

Abstract Terahertz (THz) radiation is in powerful region of electromagnetic spectrum because of prosperous application areas yet deficiency still exists about sources and detectors in despite of improvements of the research field in this range. This gap can be filled by focusing on development of THz detectors. Therefore, bolometers were preferred through many detectors due to detection sensitivity above 1 THz. In this study, $Bi_2Sr_2CaCu_2O_{8+\delta}$ (Bi2212) single crystals were used to fabricate THz bolometric detector. Bi2212 single crystals were transferred on sapphire substrate by cleavage process and e-beam lithography and ion beam etching were used to fabricate the microchip clean room facilities. Custom-designed cryogenic cryostat was used for a-b axis electrical and THz response measurements with liquid nitrogen cooled system. After electrical measurements, Bi2212 microchips detected the signals using Stefan-Boltzmann Lamp and response time were calculated. This study have shown with our experimental results that Bi2212 single crystals are potential candidates for THz bolometric detectors.

Keywords Terahertz Sensitive • Detector • Bolometer • Microchip • Bi2212 single crystal • Response Time • High-Tc Superconductor

---

T. Semerci (✉)
Department of Materials Science and Engineering, Izmir Institute of Technology, Urla, Izmir 35430, Turkey
e-mail: tugcesemerci@iyte.edu.tr

Y. Demirhan • L. Ozyuzer
Department of Physics, Izmir Institute of Technology, Urla, Izmir 35430, Turkey

N. Miyakawa
Department of Applied Physics, Tokyo University of Science, Tokyo, Japan

H.B. Wang
National Institute for Materials Science, Tsukuba 3050047, Japan

M.F. Pereira, O. Shulika (eds.), *THz for CBRN and Explosives Detection and Diagnosis*, NATO Science for Peace and Security Series B: Physics and Biophysics,
DOI 10.1007/978-94-024-1093-8_11

11.1 Introduction

By virtue of the fact that THz radiation (0.1–10 THz) has a wide range between microwaves and infrared regions in electromagnetic spectrum, noticeable applications of THz region exist, fundamentally in imaging and spectroscopy research fields [1, 2] Furthermore, THz technology is an effective area to examine the explosive materials in view of that many materials have a special peak at THz range as fingerprints [3]. According to this reason, detectors are candidates to be an important part of THz research fields via improving their sensitivities, but deficiency still exists about sources and detectors in despite of improvements of the research field in this range. This gap can be filled by focusing on the development of THz sources and detectors [4–7]. Many application areas such as imaging, spectroscopy technology will take advantage of filling this gap and developing THz sources [1]. Moreover, detectors have been especially developed to utilize THz radiation range. Recently, many detectors have been produced and developed which are based on different principles and used for distinct application areas but the common important point between these developed detectors is the increasing of the detectivity, sensitivity and decreasing of Noise Equivalent Power (NEP) which are the parameters to determine the performance of the detector.

When comparing the detectors with respect to sensitivities, cryogenic detectors such as bolometers provide more efficiency than room temperatures and also development of NEP for bolometers have been enhanced as diminished NEP by a factor 10^{11} in 70 years. Bolometer is a thermal detector of heat or power, based on measurement of the change of resistance which is a response to incoming THz radiation. Similarly, Hot Electron Bolometer (HEB) consist of two metal layers which are connected to each other with a microbridge. There is a small change of resistance occurred comprised from the incoming THz radiation through microbridge at the near transition region in HEB. This change of resistance is caused from hot electrons on the microbridge and determines the amount of incoming radiation onto the bolometer. When incoming THz radiation power P is detected by the bolometer, temperature (T_B) is increased by the ratio of $\frac{dT_B}{dt} = \frac{p}{c_{th}}$ and incoming radiation is ceased as it backs to normal state by $\tau = \frac{c_{th}(\text{heat capacity})}{c_{th}(\text{thermal conductance})}$ time constant (response time) (τ). Time constant is the remarkable parameter to determine the detectivity and sensitivity of detector which should be small (fast response time) with small heat capacity and high thermal conductance values.

Another notable parameter for getting fast response time from the bolometer is to specify the suitable material to get higher absorbance from incoming radiation with higher detectivity and sensitivity. Material of this chip is determined by the parameter, Temperature Coefficient of Resistance (TCR) $= \frac{1}{R}\frac{dR}{dT}$ which changes with respect to kind of materials. Necessary parameters are to obtaining best result for detectivity and sensitivity for detector with the large TCR and slope of the electrical measurement with small bolometer resistance.

In HEB, there is an antenna structure in which hot electrons escape from superconductor through normal metals by diffusion. Antenna structures have been successfully applied to supply the electromagnetic waves on focusing them onto microbridge by comprehend THz radiations [8, 9]. Length of microbridge is an important parameter for detection in this mechanism and should be small as much as possible to get higher sensitivity for the bolometer depending on the formula, $L_{max} = 2(D_e\tau_{ee})^{1/2}$, where τ_{ee} is the electron-electron interaction time and D_e is electron diffusion [10].

Researchers have gone towards the THz detectors, which contain superconducting materials and it has been known for 70 years but recently started to be used in technological application areas. High Temperature Superconductors (HTSs) are suitable candidates for the generation of THz radiation because of their layered structure which enables the propagation of electromagnetic wave and have higher energy gap than low temperature superconductors. Therefore, HTSs give great response at the frequency of the THz range. Bi2212 is one of the examples for high energy gap HTSs. Energy gap is wide, from 15 to 40 meV for Bi2212 which refer to wide THz gap [4, 11]. Therefore, Bi2212 is suitable for THz researches due to possession of broad energy gap. Bi2212 crystal has a critical temperature 95 K and layered structure property makes it easy to cleave the crystal during the fabrication of chip. Doping level and critical temperature of Bi2212 superconductor can be changed by adding or removing oxygen with number of Cu-O layer and affect a-b axis resistance-temperature measurement.

There are many detectors fabricated and developed from superconducting materials, especially mostly used superconductors are NbN and Nb based recently because of smaller time constants. Nb-based superconducting tunnel junction using as a THz radiation detector was performed with the range of 1–2 THz with response about 15–20 μs [12]. Later on, superconducting bolometer for THz radiation based on NbN film with log-periodic in antenna was achieved a time constant with approximately 150 ps in the range of 1–5 THz [13]. Additionally, superconducting hot-spot air bridge bolometer based on 50 nm thick Nb film with 15 μm long bridge have been demonstrated and measured by response time 1.2 μs and imaging was performed with this air bridge bolometer [14, 15]. In the year 2013, 4 mm thick NbN film based on planer logarithmic spiral antenna HEB was fabricated in the range of 0.3–3 THz and measured response time as 50 ps [16]. Terahertz detector based on $YBa_2Cu_3O_7$ superconductor microbolometer was demonstrated by Hammer et al. such that responsivity is related to size of antenna structure (1.5 μm × 1.5 μm area structure has nearly 180 V/W and 4 μm × 1 μm area structure has 100 V/W as responsivity) [17].

In this study, bolometric detector was preferred through different kinds of detectors due to detection above 1 THz frequency. As far as we know, fabricated microchip from high temperature superconducting Bi2212 single crystal for THz detector is a first study in the literature. Electrical (a-b axis) and bolometric measurements were done and time constant was calculated for THz bolometer based on Bi2212 single crystal with in this study.

11.2 Fabrication and Experimental Setup

Roughly 700 μm × 400 μm Bi2212 single crystal was glued onto 6 mm^2 sapphire substrate by fast dry epoxy in this study. The crucial part for epoxy is that it should be suitable for variation in temperature during measurements. After the crystal was glued onto substrate, it was cleaved by scotch tape to get the fresh and smooth surface. Before the cleaving process, thickness of the Bi2212 crystal is almost 700–800 nm and this thickness could be minimized until 100–200 nm after cleaving, intended thickness could be obtained by this process. Thickness of crystal is one of the significant parameters to determine the superconductivity property of the crystals [18]. Edge of transferred Bi2212 crystals were cut as pyramidal shapes and masked with aluminum foil to provide the bridge for antenna-like structure. 150–200 nm Au was thermally evaporated on the cleaved, fresh crystal surface immediately to prevent undesirable chemical reactions and any contaminations on the surface [19]. After deposition with Au layer, antenna-like structure were patterned onto crystal surface by conventional e-beam lithography and fabricated using clean room facilities and argon ion beam etching techniques. Finally, gold wires were connected on four paths in order not to cause any contact resistance. The SEM image of final outlook microchip can be seen in Fig. 11.1.

The grey area (600 μm × 1200 μm) indicates the Bi2212 crystal and sphere darker areas represent the silver epoxy in Fig. 11.1. The bright area is the antenna-like structure of microchip which has 100 μm bridge distance and 1200 μm total distance of antenna-like structure. Additionally, the design has 300 μm^2 contact paths and 450 μm distance in between them.

Bi2212 microchips were characterized in our custom-designed cryostat for electrical and bolometric measurements. This cryostat has two main chambers: First, inner part is used for cooling the microchip with liquid nitrogen (77 K) and it has 153 mm diameter, 150 mm height and 3 liters capacity. Second, microchip is vacuumed by outer chamber which has 250 mm height and 200 diameter and provides to keep in low temperatures during characterization until 30 h. Electrical measurement setup to determine the superconducting transition region and operating temperature (middle point of the operating region) of the fabricated microchip can be seen in Fig. 11.2.

According to operating temperature of the microchip, bolometric measurement was done to detect the THz radiation. Stefan Boltzmann Lamp was placed in electrical measurement setup with a distance of 5 cm from the cryostat, to determine the response of the detector. For this detection, microchip was set to the operating temperature by heater in cryostat and pulsed radiation was sent at certain times from Stefan Boltzmann Lamp to the microchip by passing through Polyethylene (PE) window which is transparent in the THz region and focusing on the wave collector. Hitting radiation on the antenna-like structure cause to increase the resistance of microchip which is observed via Labview program. This distinction was induced by increasing the temperature of the microchip during focusing of the

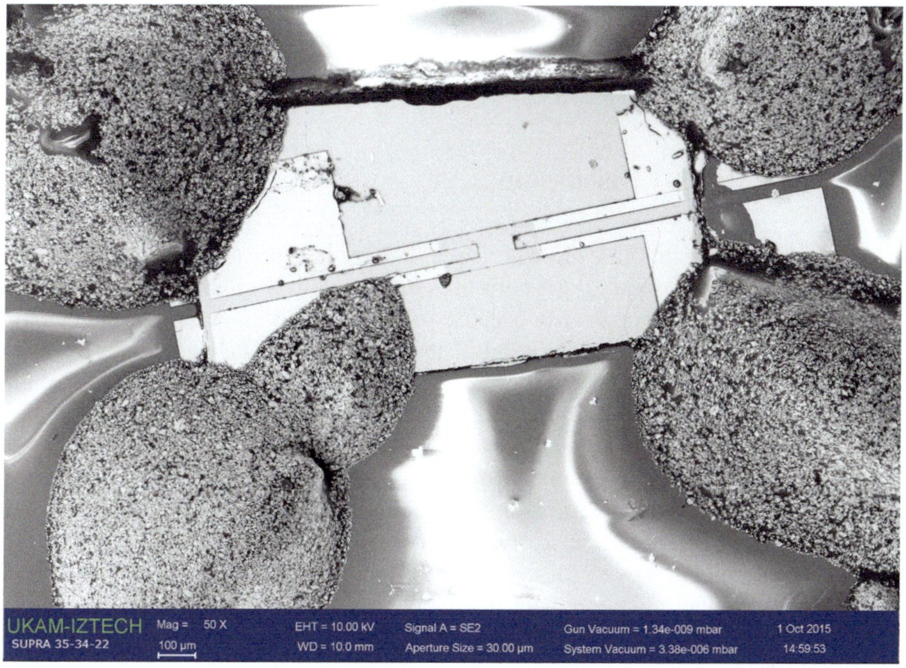

Fig. 11.1 SEM image of antenna-like patterned microchip based on Bi2212 crystal

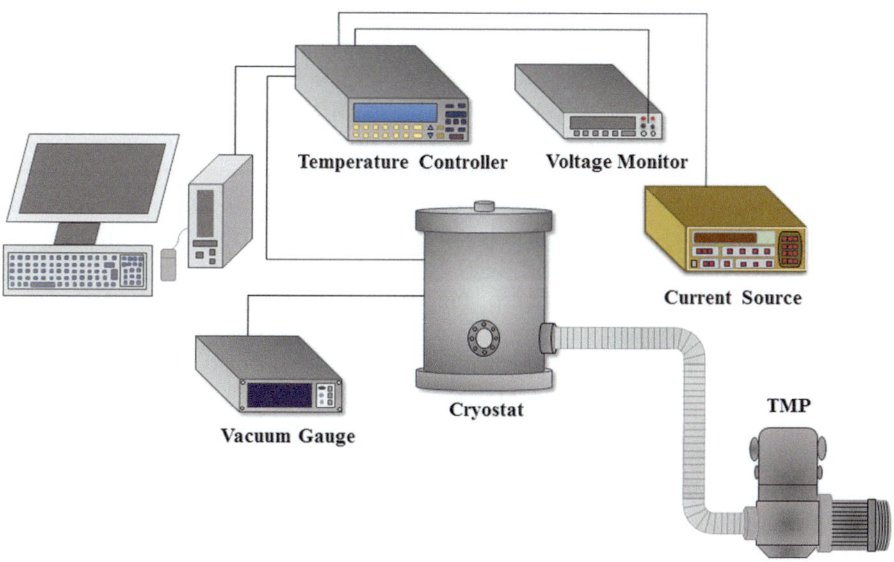

Fig. 11.2 Scheme of the electrical measurement setup

radiation. After measuring the variation of the resistance, time constant was calculated from bolometric measurement results.

11.3 Results and Discussion

Electrical characterization setup was shown in Fig. 11.2 the a-b axis of the microchips by applying constant 10 μA current is measured. It is aimed to observe sharp superconducting transition region because of the fact that this bolometric detector principle is based on the middle point of the superconducting transition region to get the highest temperature factor of the resistance. Electriccal measurements of all fabricated microchips have around the critical temperature as 90 K. Temperature dependence a-b axis resistance also establishes the onset and zero resistance of the microchips. These critical parameters depend on both oxygen level and thickness of the Bi2212 single crystals [20, 21]. The thickness of our crystals were changed between 100 and 500 nm because of the uncontrolled cleavege process. It is uneasy to arrange the thicknesses of the crystals during this process. Watanebe et al. also have shown that there is no contact resistance problem if zero resistance is below 3Ω [20]. All our microchips have been shown no contact resistance during both electrical and bolometric measurements.

Figure 11.3 shows a-b axis temperature resistance result of one of the fabricated microchip which has critical temperature (T_c) 89 K, room temperature resistance (R_{300K}) 0.54 Ω, critical temperature resistance (R_{Tc}) 0.26 Ω and variation of temperature in transition region (ΔT_c) 7.20 K. After obtaining these electrical measurement parameters for each fabricated microchip, operating temperature was selected as a middle point of the superconducting transition region and microchip was kept in that temperature via heater during bolometric measurement which was applied by switching on-off a Stefan Boltzmann Lamp at certain times. After that, response time was calculated for each microchip by indicating the initial resistance point as reference point (almost zero). Response time is the elapse time 1/e of the moment of switch off the signal.

Bolometric measurement of one of the fabricated microchip can be seen from Fig. 11.4, with an operating temperature 84 K which refer to 0.2503 Ω. Stefan Boltzmann Lamp was switched on-off at each 40 s intervals and 1.8 mΩ resistance change was observed and calculated response time value is 1.98 s for this microchip. Consequently, response time of our fabricated microchip was higher than expected values because of the fact that we had the limitation speed of the data transfer during bolometric measurement setup, this delayed data transfer may be solved by using lock-in-amplifier there by getting more exact time constants.

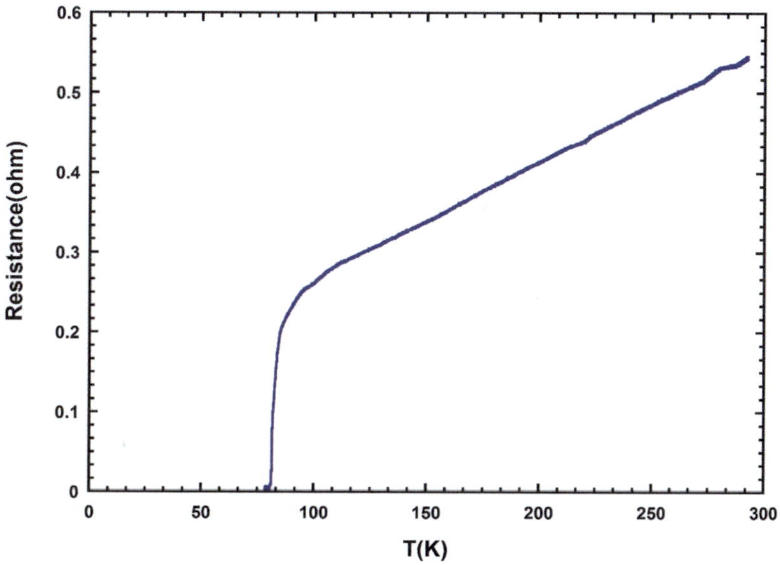

Fig. 11.3 Temperature dependence a-b axis resistance behavior of one of the fabricated Bi2212 microchip

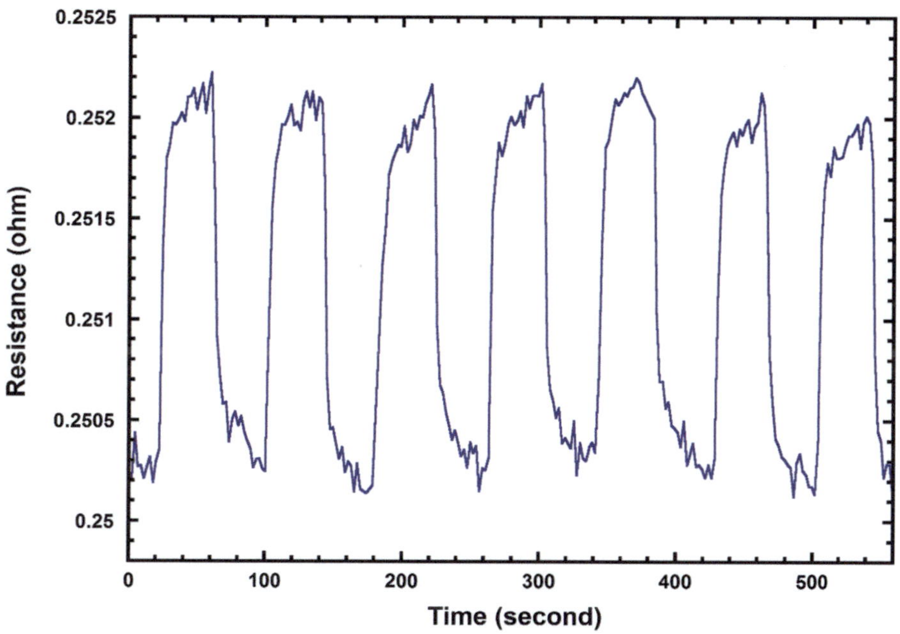

Fig. 11.4 Change of the microbridge resistance during bolometric measurement

11.4 Conclusion

In this study, we have shown that electrical and bolometric measurements of detectors which were fabricated on high quality Bi2212 single crystal in custom-designed and nitrogen cooled system. To the best of our knowledge, there is no HEB from Bi2212 single crystal in literature. Additionally, this study have shown that Bi2212 single crystals are productive candidates for bolometric detectors. Electrical measurement in a-b axis of our microchips has 90 K critical temperatures which depend on thicknesses and oxygen doping levels of the crystals. We have fabricated 100–500 nm Bi2212 thick crystals with oxygen doped range. After electrical measurements, THz radiation was detected by bolometer using Stefan Boltzmann Lamp and it has a response time 1.98 s which is higher than expected value. We concluded that this high value may be caused from speed limitation of our data transfer which might be solved by using lock-in-amplifier.

Acknowledgments This research was partially supported by Ministry of Science, Industry and Technology under project number SANTEZ 1386.STZ.2012-1 which was cooperated with ASELSAN. Additionally, we thank the Applied Quantum Research Center (UKAM) of IZTECH for the research facilities.

References

1. M. Tonouchi, Cutting-edge terahertz technology. Nat. Photon. 1(2), 97–105 (2007)
2. T. Kleine-Ostmann, T. Nagatsuma, A review on terahertz communications research. J. Infrared Millim. Tera. Wav. 32, 143–171 (2011)
3. M.C. Kemp, P.F. Taday, B.E. Cole, J.A. Cluff, A.J. Fitzgerald, W. R. Tribe, Security applications of terahertz technology, Pro. SPIE, 5070, 44–52 (2003)
4. L. Ozyuzer, A.E. Koshelev, C. Kurter, N. Gopalsami, Q. Li, M. Tachiki, U. Welp, Emission of coherent THz radiation from superconductors. Science, 318(5854), 1291–1293 (2007)
5. K. Kadowaki, M. Tsujimoto, K. Yamaki, T. Yamamoto, T. Kashiwagi, H. Minami, M. Tachiki, R.A. Klemm, Evidence for a Dual-Source Mechanism of Terahertz Radiation from Rectangular Mesas of Single Crystalline Bi2Sr2CaCu2O8+δ Intrinsic Josephson Junctions. J. Phys. Soc. Jpn. 79, 023703 (2010)
6. F. Turkoglu, L. Ozyuzer, H. Koseoglu, Y. Demirhan, S. Preu, S. Malzer, Y. Simsek, H.B. Wang, P. Muller, Emission of the THz waves from large area mesas of superconducting Bi2Sr2CaCu2O8+d by the injection of spin polarized current. Physica C Superconductivity 491, 7 (2013)
7. Y. Demirhan, H. Saglam, F. Turkoglu, H. Alaboz, L. Ozyuzer, N. Miyakawa, K. Kadowaki, Area dependence and influence of crystal inhomogeneity on superconducting properties of Bi2212 mesa structures. Vacuum 120, 89–94 (2015)
8. S. Cibella, P. Carelli, M.G. Castellano, V. Foglietti, R. Leoni, M. Ortolani, G. Torrioli, A Superconducting Bolometer Antenna-Coupled to Terahertz Waves. Journal of Low Temperature Physics 154(5–6), 142–149 (2009)
9. R.H. DuHamel, D.E. Isbell, Broadband Logarithmically Periodic Antenna Structures. IRE National Convention Record. 1, 119 (1957)
10. F. Sizov, A. Rogalski, THz Detectors-Review. Progress in Quantum Electronics. 34, 278–347 (2010)

11. L. Ozyuzer, Y. Simsek, H. Koseoglu, F. Turkoglu, C. Kurter, U. Welp, P. Muller, Terahertz wave emission from intrinsic Josephson junctions in high-T-c superconductors. Superconductor Science and Technology. 22(11) (2009)
12. C. Otani, T. Taino, R. Nakano, K. Hoshino, T. Shibuya, H. Myoren, S. Ariyoshi, H. Sato, H. M. Shimizu, S. Takada, K.Kawase. A broad-band THz radiation detector using a Nb-based superconducting tunnel junction. IEEE Trans. on Appl. Supercon. 15, 2 (2005)
13. A.D. Semenov, H. Richter, H.W. Hübers, K.S. LL'in, M. Siegel, Superconducting noise bolometer for terahertz radiation. Appl. Phys. Lett. 87, 173508 (2005)
14. S. Cibella, M. Ortolani, R. Leoni, G. Torrioli, L. Mahler, Ji-Hua Xu, A. Trediccui, H.E. Beere, D.A. Ritchie, Wide dynamic range terahertz detector pixel for active spectroscopic imaging with quantum cascade lasers. Appl. Phys. Lett. 95, 213501 (2009)
15. S. Cibella, M. Ortolani, R. Leoni, G. Torrioli, L. Mahler, Ji-Hua Xu, A. Trediccui, H.E. Beere, D.A. Ritchie, Optical characterization of superconducting hotspot air-bridge bolometer, 35th International Conference on Infrared, Milimeter and Terahertz Wave IEEE. 1–2, (2010)
16. N. Palka, P. Zagrajek, A. Czerwinski, T. Trzcinski, E. Rurka, M. Szustakowski, M. Sypek, Hot Electron Bolometer for detection of fast terahertz pulses from optical parameter oscillator. Milimetre Wave and Terahertz Sensors and Technology V Proc. of SPIE. 8544 (2013)
17. A. Hammar, S. Cherednichenko, S. Bevilacqua, V. Drakinskiy, J. Stake, Terahertz direct detection in YBa2Cu3O7 microbolometers. IEEE Transactions on Terahertz Science and Technology. 1(2), 390–394 (2011)
18. K. Yamaki, M. Tsujimoto, T. Yamamoto, A. Furukawa, T. Kashiwagi, H. Minami, K. Kadowaki, High-power terahertz electromagnetic wave emission from high-Tc superconducting Bi2Sr2CaCu2O8+d mesa structures. Optics Express. 19(4), 3193–3201 (2011)
19. I. Kakeya, K. Fukui, K. Kawamata, T. Yamamoto, K. Kadowaki, Quantum oscillation of the c-axis resistivity due to entrance of pancake vortices into micro-fabricated Bi2Sr2CaCu2O8+d intrinsic Josephson junctions. Physica C. 468, 669–673 (2008)
20. T. Watanabe, T. Fujii, A. Matsuda Anisotropic resistivities of precisely oxygen controlled single-crystal Bi2Sr2CaCu2O8+delta: Systematic study on "spin gap" effect. Physical Review Letters. 79(11), 2113–2116 (1997)
21. L.X. You, A. Yurgens, D. Winkler, C.T. Lin, B. Liang, Superconducting properties of ultrathin Bi2Sr2CaCu2O8+x single crystals. Journal of Applied Physics. 98, 033913 (2005)

Chapter 12
Integrating THz Sensors/Structures Through Electrowetting in Dielectrics (EWOD) for Security Applications

L. Sirbu, L. Mihai, M. Danila, V. Schiopu, A. Matei, F. Comanescu, A. Baracu, A. Stefan, T. Dascalu, and R. Muller

Abstract We developed a combination of technology for deposition contacts/ wires upon nanoporous InP thin film structures and RF sputtering InP films. Indium phosphide (InP) films were deposited onto glass substrate using RF magnetron sputtering by varying the substrate temperature (50–100 °C), under constant argon pressure ($6.3 \cdot 10^{-3}$Bar) and RF power (100 W). To have good optimisation of growing parameters of LT-InP films we use several characterization techniques, Raman, FTIR, XRD, and THz spectroscopy, respectively. Doping with Ga ions was used to induce disordered in InP film, to reduce the optical recombination in IR spectra.

Keywords InP film • RF sputtering • Porous InP • THz antenna • THz-spectroscopy • THz imaging

L. Sirbu (✉)
Ghitu Institute of Electronic Engineering and Nanotechnologies, 3/3 Academiei Street, Chisinau MD-2028, Moldova

School of Electrical Engineering and Computer Science Department of Physics, Max Planck Centre for Extreme and Quantum Photonics, University of Ottawa, Advanced Research Complex, 25 Templeton Street, Ottawa, ON K1N 6N5, Canada

Laboratory for Quantum Semiconductors and Photon-based BioNanotechnology, Interdisciplinary Institute for Technological Innovation (3IT), Department of Electrical and Computer Engineering, Université de Sherbrooke, 3000, boul. de l'Université, Sherbrooke QC J1K 0A5, Canada
e-mail: sirbu_lilian@yahoo.de

L. Mihai • A. Stefan • T. Dascalu
National Institute for Laser, Plasma and Radiation Physics, PO Box MG-36, Magurele 077125, Romania

M. Danila • V. Schiopu • A. Matei • F. Comanescu • A. Baracu • R. Muller
National Institute for R and D in Microtechnologies – IMT Bucharest, Erou Iancu Nicolae 126A Street, Voluntari 077190, Romania

M.F. Pereira, O. Shulika (eds.), *THz for CBRN and Explosives Detection and Diagnosis*, NATO Science for Peace and Security Series B: Physics and Biophysics, DOI 10.1007/978-94-024-1093-8_12

12.1 Introduction

The THz radiation can be produced by various methods (free electron lasers, quantum cascade lasers, Gunn diode, etc) and in different forms (continuous wave, pulsed, monochromatic, broadband). The scope of this work was to generate broadband short pulse THz radiation in photoconductive antenna [1].

A short optical pulse (fs) generates charges in a semiconductor area between two surface electrodes. These charges are accelerated in the electrical field applied between electrodes and recombining very rapidly (because of structural defects) generates broad band THz pulses into the 0.1–5THz range.

The application areas for THz technologies are rapidly expanding, and this stimulates the need for efficient THz sources. THz emission from semiconductor surfaces was investigated previously in connection with transient photocurrents that radiate THz fields as well as in the context of nonlinear optics. Up to now a low temperature grown GaAs (LT-GaAs) was successfully used for this kind of photoconductive antenna, while other materials and methods are shown to be less efficient or have not been enough studied. One of the promising materials though, as well as LT-GaAs, is LT-InP and nanostructured InP.

In analogy with many similar applications of both optical and microwave radiation, terahertz radiation can be used for imaging and sensing. The unique properties of terahertz radiation include the transparency of common packaging materials such as glass and plastics for drugs detection, as well as tissues and skin, for medical applications. The sub-millimetre wavelength permits imaging with a diffraction-limited resolution similar to that of the human eye, and many of the proposed target substances exhibit unique spectral fingerprints in the terahertz range, which can be used for identification and chemical analysis [2].

12.2 Device Fabrication

We found that nanoporous InP structure has much larger efficiency as a THz emitter and exhibit a great potential as sensitive THz detector in comparison with a bulk material. We developed the photoconductive antenna based on structured InP material and have demonstrated that irradiating nanoporous membranes of (111) InP with heavy noble-gas ions enhances terahertz emission [3].

We developed a technology for deposition contacts/wires upon nanoporous InP thin film structures, see the Figs. 12.1 and 12.2. In this way we demonstrate the possibility of fabrication an array of antennas/detectors.

The porous structure have been made using the [111]-oriented substrates of n-InP single crystals with 500 μm thickness and free electron concentration of 1.3×10^{18} cm^{-3}. The electrochemical etching was carried out in 5% HC1 aqueous solution in potentiostatic regime as described elsewhere [4].

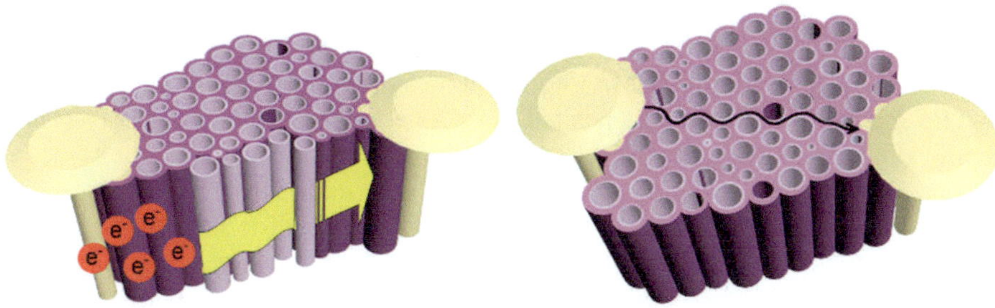

Fig. 12.1 Carrier density of nanoporous InP is represented by violet color, depletion layer by grey and with solid arrow/line show a conductive pathway

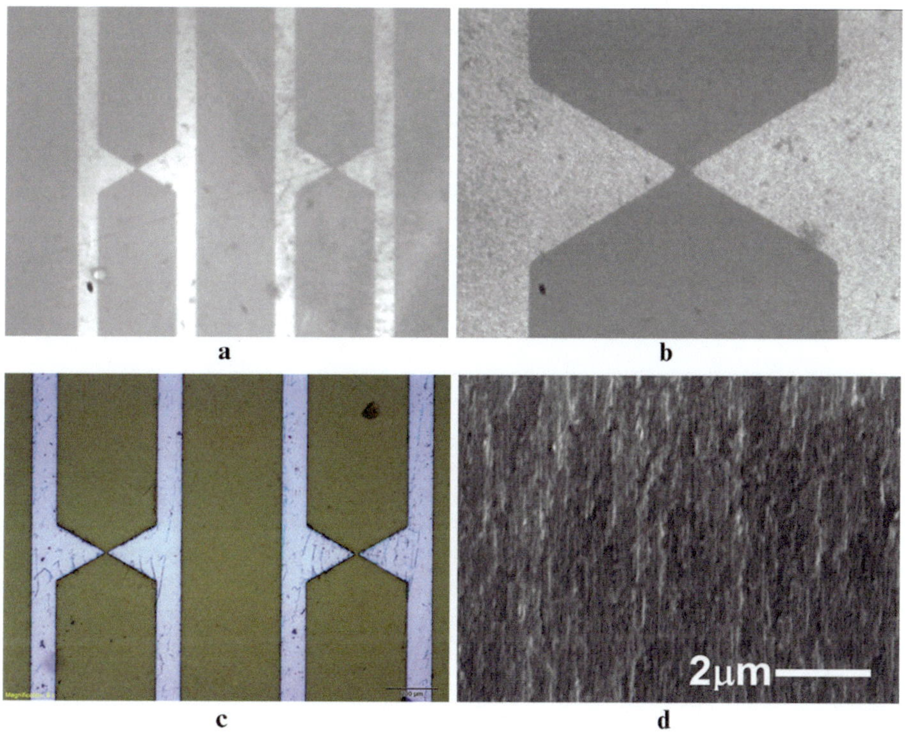

Fig. 12.2 Optical image of THz antenna on porous InP (**a**, **b**), on InP film (LT-InP) (**c**), and SEM image of cross section of porous InP filled with Au (**d**)

Another method for fabrication of photoconductive antenna is RF deposition of LT-InP. Indium phosphide (InP) films were deposited onto glass substrate (later, on sapphire) with the thickness of film 100–2000 nm, using RF magnetron sputtering by varying the substrate temperature (50–100 °C), under constant argon pressure and RF power (40–100 W). Some of the InP thin film samples were annealed at 450 °C.

A positive photoresist (PMMA) was used for both type of InP samples (porous, and RF magnetron sputtering deposited films) to configure Cr-Au contact pads and dipole antennas, obtained by liftoff method.

The elemental composition of the films was determined by electron dispersive X-ray analysis using EDX, which consist of ~40%atoms of P and ~60%atoms of In. The similar results are presented in reference [5, 6]. In order to increase the stoichiometry ratio (~50%atoms of P and ~50%atoms of In) we use two methods:

1. Changing the growing parameters for the RF sputtering deposition, and
2. Selective etching of In [7].

The InP thin film properties were investigated using different techniques such as XRD, Raman, FTIR and THz spectroscopy and imaging. The XRD measurement was performed with an X-ray diffractometer SmartLab Rigaku (IMT-Bucharest, RO) employing Cu Ka radiation from a rotating anode source and a Ge monochromator.

The Raman spectra were recorded by a Renishaw spectrometer using a 785 nm diode laser as excitation source.

The spectral investigations in THz area (ranging from 0.06 up to 3 THz) was made using a TPS3000 THz spectrometer (at INFLPR – Magurele, RO), using the transmittance module. The THz imaging was obtained using RIM module, from the same equipment, using a scanning step of 50 μm for the entire sample surface.

12.3 Results

The results obtained from X-ray diffraction are represented in Fig. 12.3a–d. We can see that by decreasing the sample growth temperature the intensity of the In peak is decreasing (Fig. 12.3a–c) and that the thin film obtained through this method remained amorphous (small crystallites ~1.8 nm). The best result was observed in the case of the selectively etched samples (Fig. 12.3d), where X-ray diffraction spectra show well resolved InP peaks. Also, as we can see from the Fig. 12.3a–c, that the peaks for InP look broader than normally found for bulk material, which is typical for small crystallites (the size is estimated by Williamson-Hall method, and it consist ~1.8 nm for samples from Fig. 12.3a–c and ~20 nm for samples from Fig. 12.3d). The indium peak is decreasing with temperature of film deposition.

In the IR spectrum of the InP thin film (Fig. 12.4) can be noticed a stretching vibration peak of C-H hydroxyl group (the functional group in alcohols (ethanol, methanol, etc.)) at 2906 cm^{-1}, in addition, there are peaks for In-H at 1356 cm^{-1}, 1546 and 757 cm^{-1}, P-H at 836 cm^{-1}. There are few changes after annealing in Ar atmosphere for 30 min at 450 °C (see Fig. 12.4, the normalized spectrum): C-H at 2882 cm^{-1}, In-H at 1397 cm^{-1}, 1542 and 686 cm^{-1}, P-H 855 and 686 cm^{-1}. We

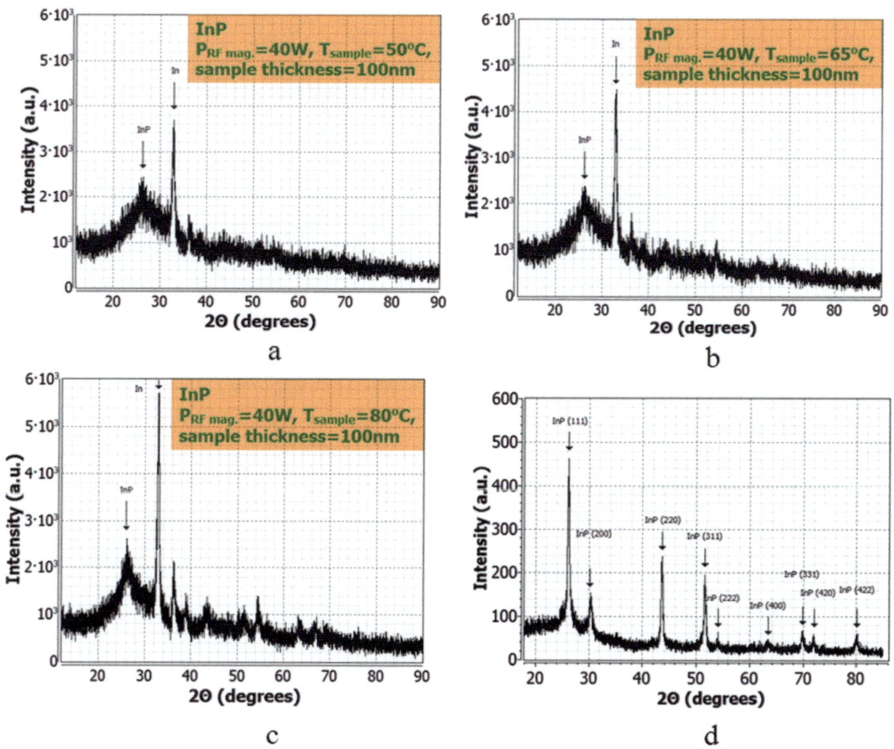

Fig. 12.3 XRD pattern for InP thin films

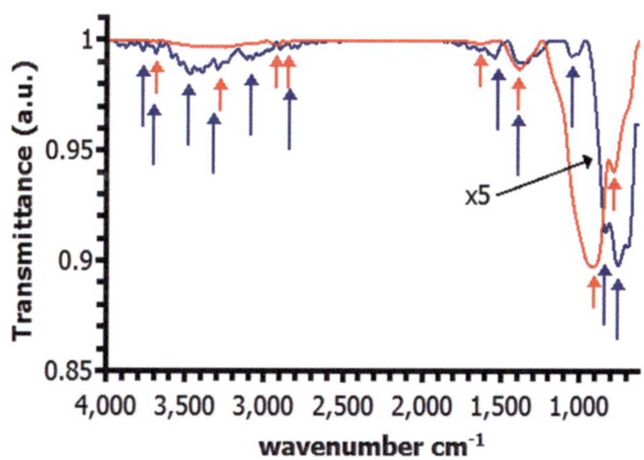

Fig. 12.4 FTIR spectra of InP thin film. The normalized spectrum of the InP annealed in Ar atmosphere for 30 min at 450 °C is multiplied by factor 5

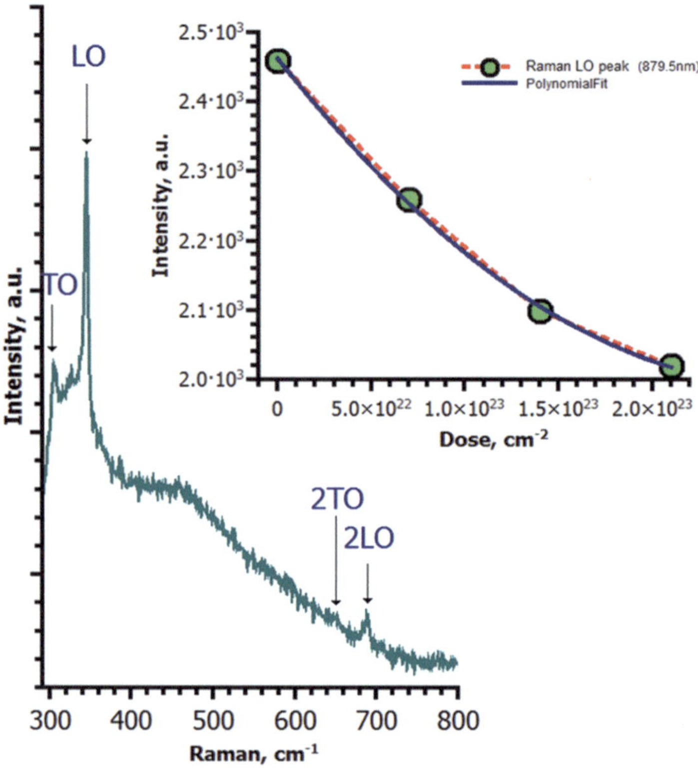

Fig. 12.5 Raman spectrum, inset is presented the Raman intensity vs. density of implanted of Ga ions (measurements were supported by the FP7 project MOLD-ERA (Grant no 266515), Bristol, UK)

detected the peaks 1059 and 1025 cm^{-1}, respectively, related to the $InPO_4$ oxide formation after electrochemical processes [8].

The Raman spectrum of InP thin film is presented in Fig. 12.5. The two relatively sharp peaks around 307 and 343 cm^{-1} were assigned to the first-order scattering from TO and LO phonons of InP crystal, respectively, while weaker peaks at around 651 and 688 cm^{-1} were assigned to the second-order Raman scattering [8]. The observed strong and narrow LO phonon peak of InP is similar to that of the bulk and manifest a good crystalline quality of obtained films. The inset of Fig. 12.5 represents the Raman LO peak intensity at different density of implanted of Ga ions, without implantation, $7.0 \cdot e^{22}$, $1.4 \cdot e^{23}$ and $2.1 \cdot e$- cm^{-2}, respectively. Doping with Ga ions was used to induce disorder in InP film, for reduce optical recombination in IR spectra. The Raman spectrum of the InP film, and we found a good correlation with literature [9, 10].

Optical pulses from a Femtolaser Fusion 20 pulsed laser at central wavelength of 805 nm having an average power of 180 mW were used to generate THz radiation.

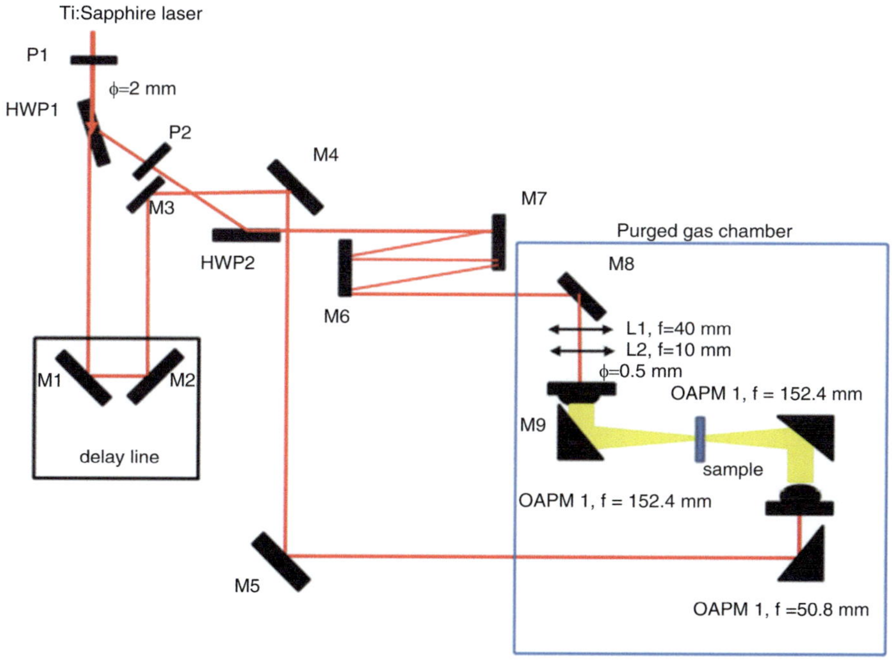

Fig. 12.6 Experimental setup for investigating THz emission from InP

The laser system delivers 20 fs pulses at 75 MHz repetition rate. The polarizer P1 and the half wave plate HWP1 split the laser beam in two arms. The most powerful one is used to pump the THz emitter and the other functions as a gate signal for the detector. The power on THz detector could be adjusted by changing the wave plate angle. The beam is directed to an optical delay line and then guided by mirrors M3, M4 and M5 to the off-axis parabolic mirror which focuses the beam onto the antenna gap. The half wave plate HWP2 and the polarizer P2 were used to adjust the power of the pump beam on THz emitter. The beam dispersion is compensated by a dispersion compensator consisting of mirrors M6 and M7. The telescope formed by the lenses L1 and L2 provides a beam diameter of 0.5 mm enhancing the power density of the collimated laser beam.

The THz emitter, an inter-digital photoconductive THz antenna with micro lens array and collimating silicon lens, delivers a collimated THz radiation which is focused on the sample by an off-axis parabolic mirror and then recollimated by a second off-axis parabolic mirror on another photoconductive antenna that was used as detector, parallel line photoconductive antenna PCA 40-05-10 from BATOP Optoelectronics. Thus, the wave form of the electric field of the THz wave can be measured by scanning the optical delay stage (see Fig. 12.6).

From the Fig. 12.7, the impedance is changing at varying the incident beam excitation fluencies linearly for porous structures. The porous InP has demonstrated

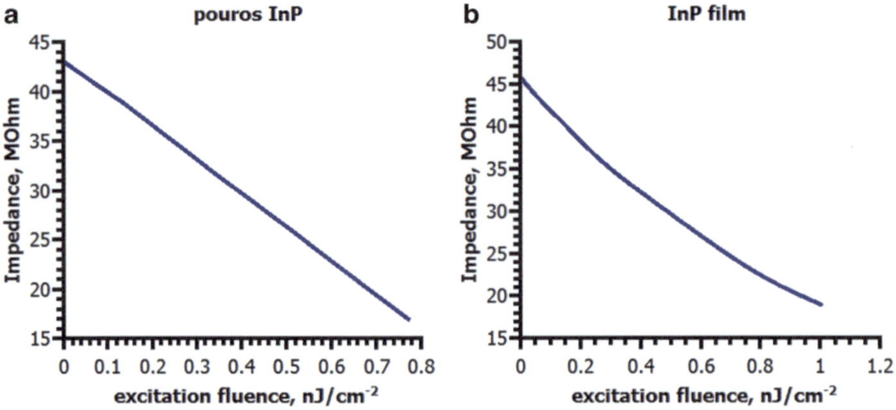

Fig. 12.7 Impedance measurements at different excitation fluencies for (**a**) porous InP and (**b**) InP film

linear behavior of decreasing sample impedance at increasing of incidence excitation fluency.

The absorption coefficient versus photon energy for InP films were measured at 300 K for three type of thickness, 500, 1000 and 2000 nm, respectively before and after HNO_3 etching. In Fig. 12.8a–c is represented the THz-TDS spectra for three different thicknesses, and insets represents the changes in amplitude of THz pulse for etched vs. initial samples.

According of early report of Aspnes and Studna [1983] the thickness of InP films were optimized for the absorption coefficient α [11].

As we can see from Fig. 12.9 the slope of the absorption coefficient vs. wavelength is changed after etching of indium layer remained after RF deposition. The power of IR lamp was ~180 (177) mW, and optical pulses from a Femtolaser Fusion 20 pulsed laser at central wavelength of 805 nm have an average power of 180 mW. We report in this paper the surface impedance measurement of a magnetron RF sputtering InP films to be 15 MΩ/cm vs PCA 40-05-10-800-x from BATOP Optoelectronics one with 9 MΩ/cm [12]. By adding 1%, and 5% of arsenium, we may change significant the impedance of InP film. Analyzing the Fig. 12.9a, b, we may conclude that the thickness of LT-InP for maximum efficiency of THz conversion is ~1000 nm. This is based on two parameters: maximum absorption rate of fs pulse with thickness of material, Fig. 12.9a, and attenuation of obtained THz radiation at exit from InP film, Fig. 12.9a–c. Raman spectra of LT-InP:As 1%, and 5% has no evidence in changing positions of LO and TO peaks. The InP film band gap is ~1.34 [10], 1.19 and 1.18 eV for As doping level of 0%, 1%, and 5%, respectively

A set of InP film prepared in different conditions (see the legend of Fig. 12.10) had been studied with the THz transmission and it had been compared with the empty holder (air) (see Fig. 12.10). From Fourier transform of THz time domain we easily can see that InP films has the bandwidth of transmitted signal ~2 THz. We suppose that the main impact of absorption is due to presence of considerable

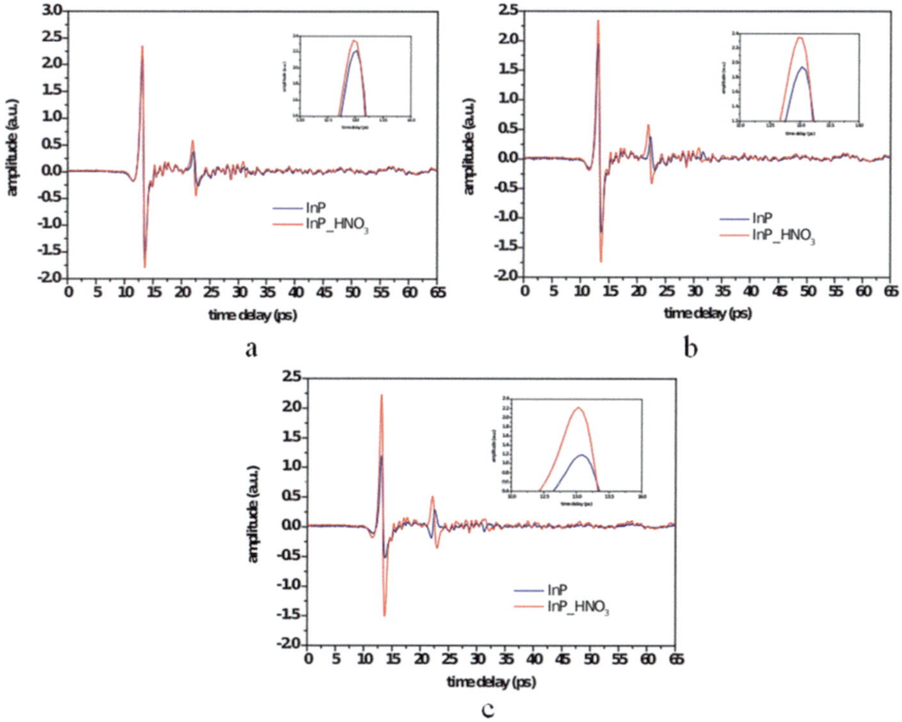

Fig. 12.8 THz TDS spectra for InP films in comparison with etched InP films with thickness of (**a**) 500 nm, (**b**) 1000 nm, and (**c**) 2000 nm respective

amount of glass alloys that has the thickness of 150 μ. Another contribution of absorption is water vapor, that because the experiments have been done in atmosphere without using nitrogen gas chamber.

Since the built EWOD structure is a transparent one for THz radiation, it allows scanning of fluid slipping through micro channel and may be applied for CBRN and explosives detection. The structure is obtained by thermal oxidation of a (100) non doped Si wafer. The grown SiO_2 on top is 500 μm thick. A positive photoresist (PMMA) was used to configure Cr-Au pads, obtained by lift off method. In Fig. 12.11a, is represented the scanned section of chip. The central part of the Fig. 12.11a, indicated by a green axes, is pointing to the pad with resonator shape in Fig. 12.11b. The pad-resonator may have other configurations, [1], for square one, and it THz spectra can be seen in Fig. 12.11c, where the signal between 1 and 2 THz corresponds to resonator scanned area.

A chip shape for the pads has been optimized in order to have a smooth moving of separate liquid micro droplet through micro channel. The applied bias depends also on width of pads and of the inter distance between them, which was estimated in Coventor software by using BabbleDrop module [1].

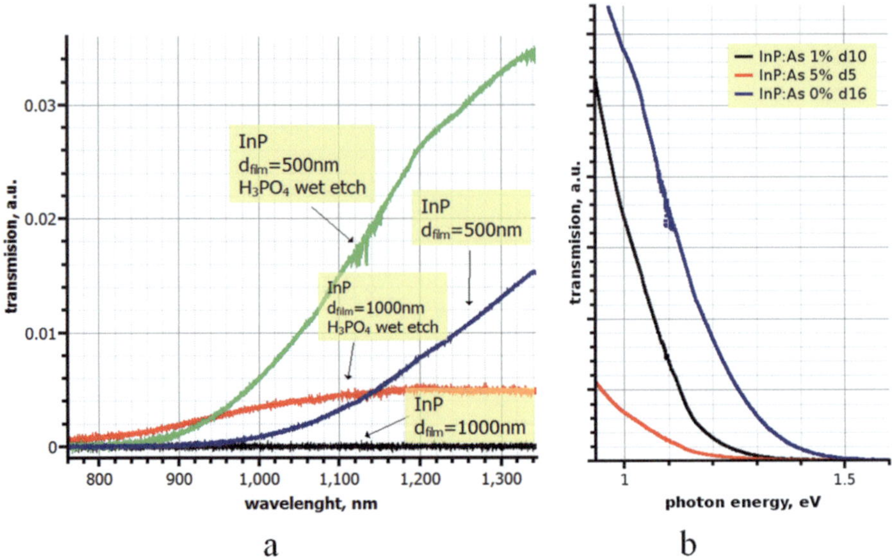

Fig. 12.9 IR spectra for InP films corresponding to thickness of 500 and 1000 nm before and after HNO_3 etching, (**b**) IR spectra for InP films which correspond with doping level of As 0%, 1% and 5%

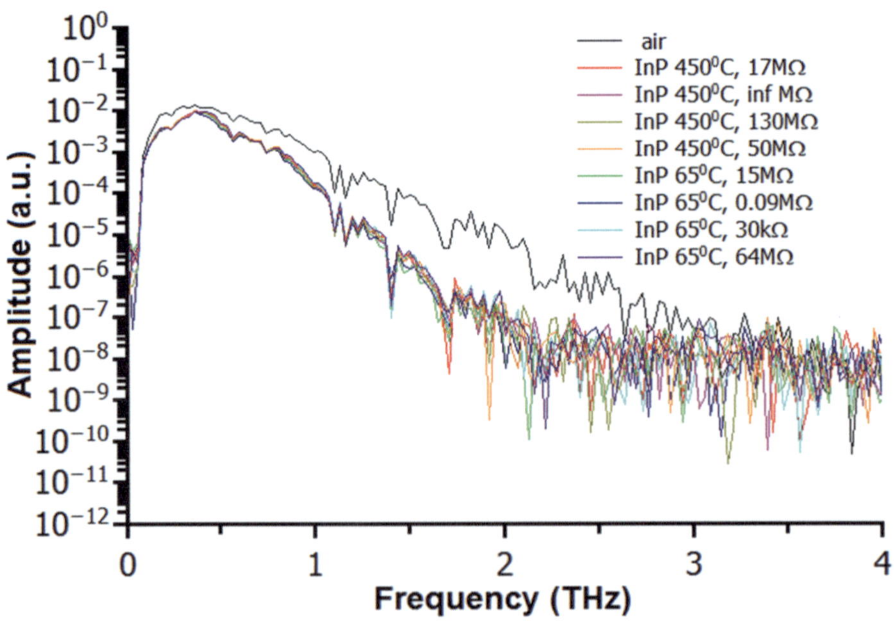

Fig. 12.10 THz spectra from InP thin films deposited on glass

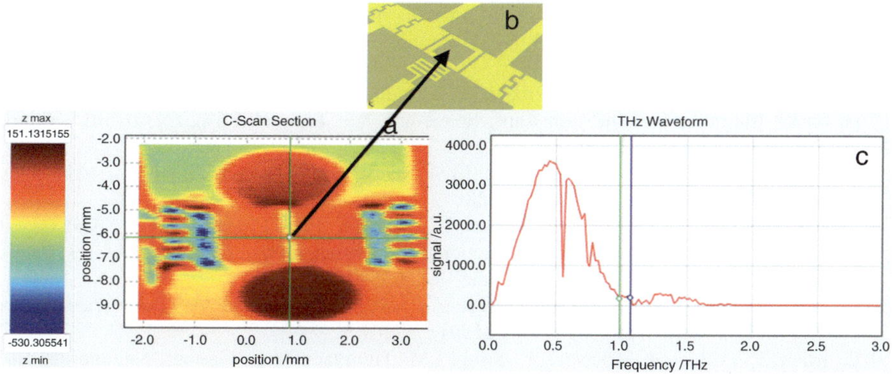

Fig. 12.11 (**a**) 2D THz mapping, (**b**) THz resonator configuration, and (**c**) THz spectra of pad-resonator

12.4 Conclusion

In this paper we demonstrated the possibility to fabricate a device that is able to handle with liquids, and to measure the THz spectra in real time. We managed to dope the InP films with As, 1% and 5% and we implanted different dose of Ga ions. The selective etching improved considerably the quality of InP films. After data analysing we may conclude that the thickness of LT-InP for maximum efficiency of THz conversion is ~1000 nm. However, the THz emission from LT-InP films was not established yet, suggesting us the dipole disorientation in such material in comparison with porous membranes reported early [3]. Doping with Ga ions was used to induce disorder in InP film, to reduce the optical recombination in IR spectra.

Acknowledgements This work was supported by The NATO Science for Peace and Security Programme.

References

1. www.batop.com/information/PCA_infos.html
2. Lilian Sirbu, Lidia Ghimpu, Raluca Müller, Irina Vodă, Ion Tiginyanu, Veaceslav Ursaki, Traian Dascalu Hydrophobic ZnO used in EWOD technology and SAW devices for better bio-fluid slip at microchannel walls controlled by DC pulses, CAS2012, Proceedings, Vol.I, 15–17 oct 2012, Sinaia, Romania, p. 231–234
3. K. Radhanpura, S. Hargreaves, R. A. Lewis, L. Sirbu, and I. M. Tiginyanu, Heavy noble gas "Kr, Xe" irradiated (111) InP nanoporous honeycomb membranes with enhanced ultrafast all-optical terahertz emission, Applied Physics Letters 97, 181921, 2010
4. S. Langa, M. Christophersen, J. Carstensen, I. M. Tiginyanu, H. Föll, "Single crystalline 2D porous arrays obtained by self organization in n-InP" Physica Status Solidi (a) Volume 197 Issue 1, Pages 77 – 82, 23 Apr 2003

5. G. Hema Chandra, J. Perez de la Cruz, Microstructural and optoelectronic properties of polycrystalline InP films deposited by RF magnetron sputtering, Journal of Crystal Growth 354 (2012) 67–75
6. G. Hema Chandra, J. Perez de la Cruz and J. Ventura, The influence of argon pressure and RF power on the growth of InP thin films, Semicond. Sci. Technol. 26 (2011) 075017. (8pp), doi:10.1088/0268-1242/26/7/075017
7. A.R. Clawson, Guide to references on III–V semiconductor chemical etching, Materials Science and Engineering, 31 (2001) 1–438
8. Olivier Pluchery, Joseph Eng Jr., Robert L. Opila, Yves J. Chabal, Etching And Oxide Formation On Inp Surfaces, Tenth International Conference on Vobrations at Surfaces, Palais du Grand Large, Saint Malo (France), June 17–21 2001
9. G. Irmer, Raman spectra of InP, in: Properties of indium phosphide, EMIS Datareview Series No.6, INSPEC: London and New York (1991), p. 207
10. G. Irmer, P. Prunici, J. Monecke, L. Sirbu, I.M. Tiginyanu & G. Gärtner. Nanoporous InP studied by micro-Raman scattering. 19th Int. Conf. on Raman Spectroscopy (ICORS-2004), Gold Coast, Australia, August 8–13, 2004.
11. Aspnes, D. E. and A. A. Studna, Phys. Rev. B27, 2 (1983) 985–1009
12. Datasheet for BATOP photoconductive antenna Available: www.batop.com/products/terahertz/photoconductive-antenna/data-sheet/manual_PCA-40-05-10-800.pdf

Part II
Materials Properties

Chapter 13
The Critical Effect of Hydration on the THz Signatures of Biomolecules and Bioparticles

E.R. Brown and W.-D. Zhang

Abstract Here we present an original study of the effect of hydration on THz absorption signatures in biomolecules (lactose monohydrate and biotin) and bioparticles (Bacillus thuringiensis and Bacillus cereus spores). We observe a "red-shift" in center frequency with increasing hydration in all samples, consistent with Lorentzian-oscillator behavior. But the effect of hydration on linewidth is ambiguous, sometimes increasing the linewidth (consistent with Lorentzian behavior) and sometimes decreasing the linewidth.

Keywords THz radiation • THz spectroscopy • Absorption signatures • Center frequency • Linewidth • Lorentzian model • Molecular crystals • Lactose • Biotin • Biomolecules • Bioparticles • Genus bacillus spores

13.1 Introduction

Terahertz (THz) radiation lies in the range 300–10,000 GHz, which currently is beyond the upper-frequency limit of wireless-communications and radar systems worldwide. It has been investigated for a variety of applications such as radio astronomy [1], security imaging [2], materials characterization [3], and medical diagnostics [4]. And of course, these applications benefit from the continual evolution of solid-state electronics and electromagnetics, such as resonant-tunneling oscillators up to 1.2 THz [5], amplifiers up to 1.0 THz [6], and frequency-multipliers and mixers up to roughly 2.0 THz [7]. Phenomenologically, MMW and THz radiation oscillate with a period comparable to the ps-timescale of collective vibrations in biomolecules and bioparticles, and also to the characteristic orientation times of dipolar liquid-water molecules. Because of this both are absorbed strongly by soft tissue of all sorts, but the collective interaction of molecules and particles can be more resonant with respect to excitation frequency. In principle, this makes it possible to distinguish the biomolecular excitation from

E.R. Brown (✉) • W.-D. Zhang
Department of Physics, Wright State University, Dayton, OH 45435, USA
e-mail: elliott.brown@wright.edu

M.F. Pereira, O. Shulika (eds.), *THz for CBRN and Explosives Detection and Diagnosis*, NATO Science for Peace and Security Series B: Physics and Biophysics,
DOI 10.1007/978-94-024-1093-8_13

water absorption, which generally causes heating. Heating in biomaterials is complicated and potentially confusing since it creates its own strong bioeffeccts.

For many years THz spectroscopy has been touted as a complement to FTIR and microwave vapor-phase spectroscopies, providing information about vibrational resonances from the tertiary structure of biomolecules and bioparticles rather than their interatomic vibrations or rotations. The challenge in THz spectroscopy has been achieving adequate selectivity and specificity needed to distinguish these signatures from various forms of "clutter", such as standing waves in sample holders and scattering effects (e.g., speckle) in grainy materials. Another challenge is measuring biosignatures "in-situ", meaning in hydrated media for biosamples. The strong absorption from liquid water makes spectroscopy virtually impossible for most living organisms *in vivo*, and even most aqueous solutions. The lack of powerful THz sources (i.e., average output $\geq$1 mW) exacerbates the problem since traditional methods for clutter mitigation, such as the placement of attenuators in the signal path to suppress standing waves, tend to reduce the THz signal-to-noise ratio deleteriously.

In this work we show that the hydration level for THz signatures is "critical", meaning that it must lie in a certain range for signatures to be readily measured. For polar molecules having a strong enough vibrational resonance, such as lactose monohydrate, the lower end of the range extends to zero and the upper end to roughly 20% hydration by mass. For non-polar molecules or bioparticles, such as Bacillus spores, the lower end of the range rises above zero to roughly 10% hydration, but the upper end remains at about 20%. In addition, hydration usually causes a reduction (i.e., "red shift") in the signature center frequency which will be described later in the document.

13.2 Experimental Methods and Samples

The spectroscopic experiments were carried out with a high-resolution frequency-domain, coherent (photomixing) spectrometer like those described in References [8, 9]. The spectra were measured in stepped-frequency mode at 1.0 GHz steps and 0.1 s integration time. The typical background spectra and noise floor are shown in Fig. 13.1b where the tuning bandwidth is seen to be at least 1.7 THz, and the dynamic range is $\approx$80 dB at the low end of the scan (200 GHz) and $\approx$40 dB at the high end (1.9 THz). The frequency range is determined mostly by the limited temperature tuning of the DFB diode lasers used in the spectrometer. For the present experiments, the frequency resolution is determined by the 1.0 GHz steps, although the instantaneous THz tone is much narrower than this.

For the present study we chose two biomolecular samples and two bioparticle samples, all known from previous experience to have strong signatures in the THz region and within the operating range of our photomixing spectrometer. The biomolecular samples were lactose monohydrate and biotin, each having the chemical compositions shown in Fig. 13.2a, b, respectively. The former is a crystallized

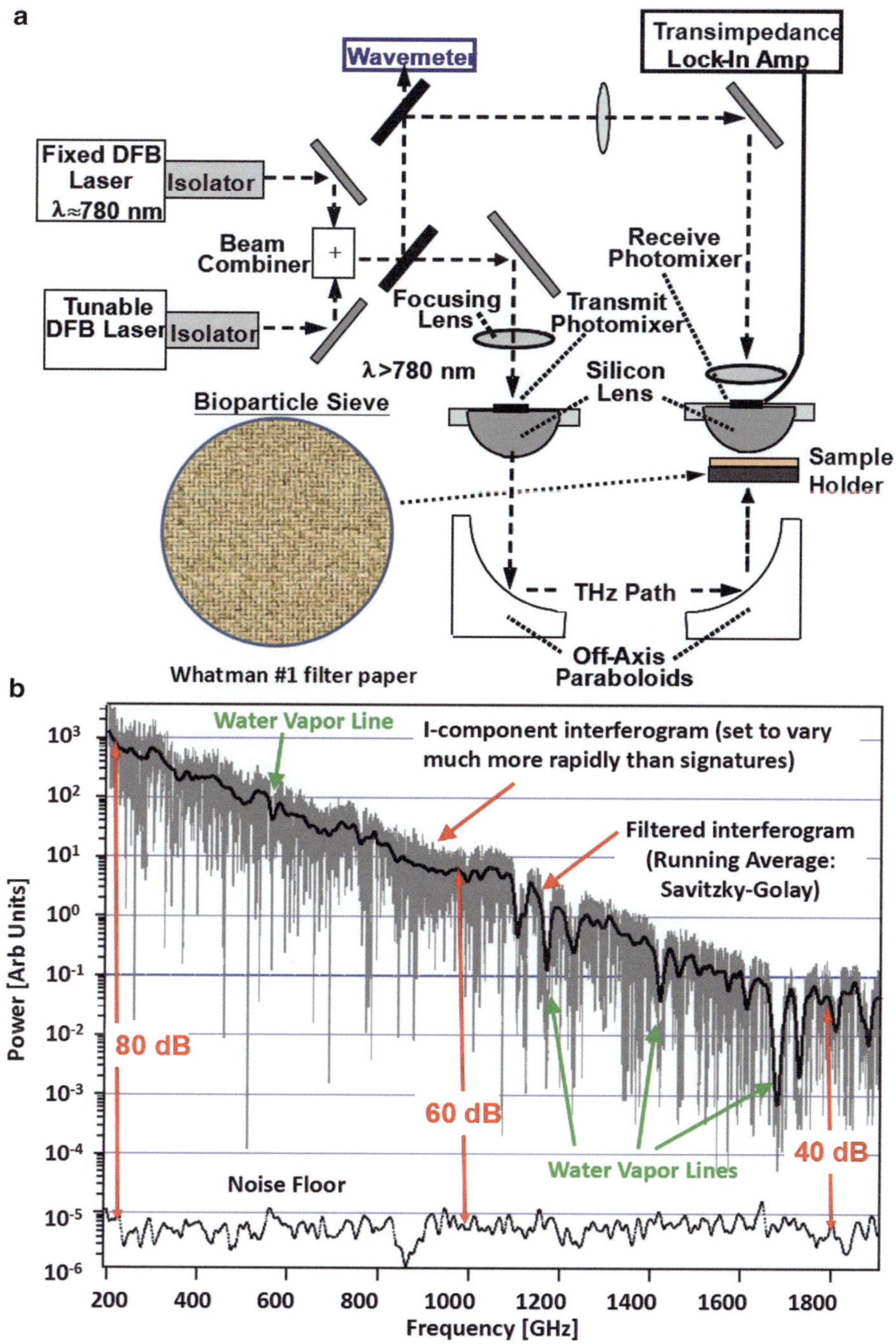

Fig. 13.1 (**a**) Photomixing spectrometer block diagram and location of biosamples, such as the Bacillus spores embedded in filter paper. (**b**) Typical performance of spectrometer from ~0.2 to 1.9 THz

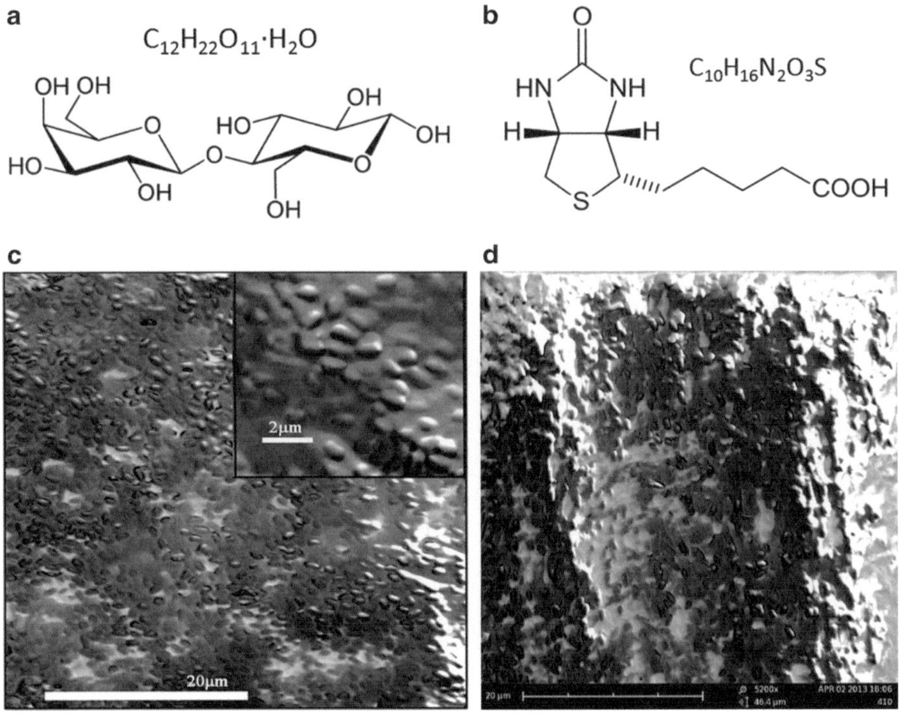

Fig. 13.2 Structural formula for lactose monohydrate (**a**) and biotin (**b**). (**c**) SEM for the B. thuringiensis (Bt) sample. (**d**) SEM for the *B. cereus* (Bc) sample. The *white horizontal line* at the *bottom left* of both photos is the 20-um scale bar. The blow-up embedded in 2(**c**) shows the telltale ~1.2-um-long ellipsoidal shape of the spores

form of milk sugar (monoclinic structure), and the latter is crystallized vitamin B7 (rhombohedral structure). Both molecular crystals have built-in polarity, are hydrophylic, and appear macroscopically as white powders with submicron grains. The lactose and biotin samples were mounted in metallic rings and encapsulated with thin (~12 um) polycellulose windows to slow the rate of evaporation. Two identical samples were made for each biomolecular type and loaded to the same approximate level of hydration, starting at between 25 and 30% *gravimetric* (accurate to $\approx$0.1%). One of the two was left on a precision balance, and the other mounted in the THz spectrometer. Then THz spectra were taken roughly every 15 min, corresponding to a typical drop of hydration on the balance-sample of $\approx$2% per spectrum.

The bioparticle samples were endospores of two Bacillus species: B. thuringiensis (Bt), subspecies kurstaki, and *B. cereus* (Bc). Bt is a common soil bacteria and insect pathogen [10, 11]. Upon sporulation, this bacteria produces bi-pyramidal protein crystals, δ-endotoxins or Cry proteins which are toxic to a variety of insects but are not harmful to humans. While Cry toxins have been widely applied as a "green" pesticide, recent use introduces Cry proteins into transgenic crops, providing a more targeted approach to insect management. The endospore and crystal are simultaneously present after an environmental stress induces sporulation in the vegetative cell, but are typically released after cell lysis. The spores can endure harsh environments such as aridity, extreme cold and heat and some radioactivity. When nutrition, temperature and other environmental conditions become favorable, spores germinate, develop into rapidly growing vegetative cells and begin another life cycle [12]. More relevant to biosensing and the CBRNE underlying theme of this NATO Workshop, Bt and Bc are in the same genus as B. anthracis [13], the notorious "Anthrax" bio-toxin. So the identification of THz signatures belonging to Bt or Bc could provide impetus for an important application of THz spectroscopic sensing, but without the handling hazards of standard biosensing techniques.

The Bacillus spores were cultured from isolated stock bacteria in 50 ml tryptic soy broth (TSB; Becton, Dickinson, Sparks, MD) at 30 C. Bacterial cultures were transferred to a test tube and centrifuged at 5000·g for 30 min. The liquid broth was removed, and then the bacteria were maintained at 4 C in the sealed test tube. Just prior to measurement, a concentrated, 0.2 mL paste-like sample of Bt or Bc bacteria (without broth) was removed from the sealed tube, and spread evenly with a spatula over $\approx$ 2 cm diameter on the filter paper. The thickness of the resulting samples ranged from 200 to 600 um. Scanning electron micrographs (SEMs) for the Bt and Bc pastes are shown in Fig. 13.2c, d, respectively, each showing a large fraction of spores as proven by the zoom-in of Fig. 13.2c.

Three samples of Bt were tested for the present study: sample Bt1 containing vegetative cells but negligible spores; sample Bt2 containing a mixture of vegetative cells and spores at low hydration; and sample Bt3 containing vegetative cells and spores at modest hydration. Also three samples of Bc were tested: sample Bc1 containing a mixture of vegetative spores at a hydration level comparable to Bt3; sample Bc2 the same as Bc1 but dried out thoroughly to have negligible hydration; and sample Bc3, the same as Bc2 but re-hydrated to a saturated level ("muddy" texture). Because of our lacking knowledge of the broth compositions and the degree of centrifuge concentration, the hydration levels are only qualitative and based on visual observation.

For all samples the spectrometric procedure was as follows. First, the THz beam path was blocked completely to obtain the noise floor spectrum $P_N(\nu)$. Then a blank sample holder was added to the beam path: an empty encapsulated ring for the biomolecular sample, or bare filter paper for the bioparticle sample. In both cases this yielded the background spectrum $P_B(\nu)$. Then the real samples were added to the beam path to yield the sample spectrum $P_S(\nu)$. Finally, the transmittance was calculated as $T(\nu) = [P_S(\nu)\text{-}P_N(\nu)]/[P_B(\nu)\text{-}P_N(\nu)]$. For the biomolecular samples, the time between successive scans corresponding to different hydration levels, was approximately 10 min.

13.3 Experimental Results

13.3.1 Biomolecules

The experimental transmission results for lactose monohydrate and biotin are displayed in Fig. 13.3a, b, respectively. The "dry" spectrum of lactose monohydrate is in good agreement with previous, high-resolution experiments and signature analysis, showing that the signature is well fit by a Lorentzian model [14]. In the dry state of Fig. 13.3a, the signature center frequency is 530 GHz and the fullwidth at half-maximum (FWHM) is $\approx$27 GHz. But when the hydration increases above $\approx$10%, the center frequency "red-shifts" significantly. The maximum shift is $\approx$4.0 GHz at 26% hydration: well-resolved by the coherent frequency-domain (photomixing) spectrometer used in this study.

For the biotin sample, the dry center frequency is $\approx$540 GHz with FWHM of $\approx$ 55 GHz - much broader than the lactose [14]. The maximum shift is $\approx$16 GHz at a hydration of 22%, and the shifted spectrum is significantly broadened compared to the "dry" spectrum – by approximately 50%. Upon further inspection, it appears that beyond 20% hydration in Fig. 13.3b, the signature may be transforming into a "doublet", and beyond 25% may be saturating at the noise floor of the instrument. This exemplifies a key problem with THz biospectroscopy, which is that the aqueous local environment is often critical. That is, water in moderate concentrations generally strengthens the THz signatures, particularly by inducing a polarity if one doesn't already exist or is weak. But at larger concentrations the water masks the THz signatures at least through Debye absorption. This is evident in the lactose data of Fig. 13.3a at 29.0% hydration, and in the biotin data of Fig. 13.3b at 25.5%, where the transmittance spectra saturate at the signature-center frequency because the transmitted signal drops into the instrumental noise floor.

13.3.2 Bioparticles

The transmission spectra of samples Bt1, Bt2, and Bt3 are plotted in Figs. 13.4a–c, respectively. Sample Bt1 having negligible sporulation and low hydration, shows a relatively high transmission but no obvious signatures except for an anomalous one near 700 GHz. Sample Bt2 having significant sporulation and also low hydration, shows four distinct signatures centered at 966, 1026, 1098, and 1164 GHz. The last two coincide with well-known and strong water-vapor absorption lines [15], but the others do not. And because Bt2 has spores but Bt1 is lacking, the 966 and 1026 signatures are attributed to the spores [16] and labelled α and β, respectively, in Fig. 13.4b. There are also two broader, weaker features centered around 480 and 660 GHz that are not near any water vapor lines, and are addressed further below.

The transmission spectrum of samples Bt3 in Fig. 13.4c shows six distinct signatures centered at 556, 752, 955, 1015, 1098, and 1164. The last two are the

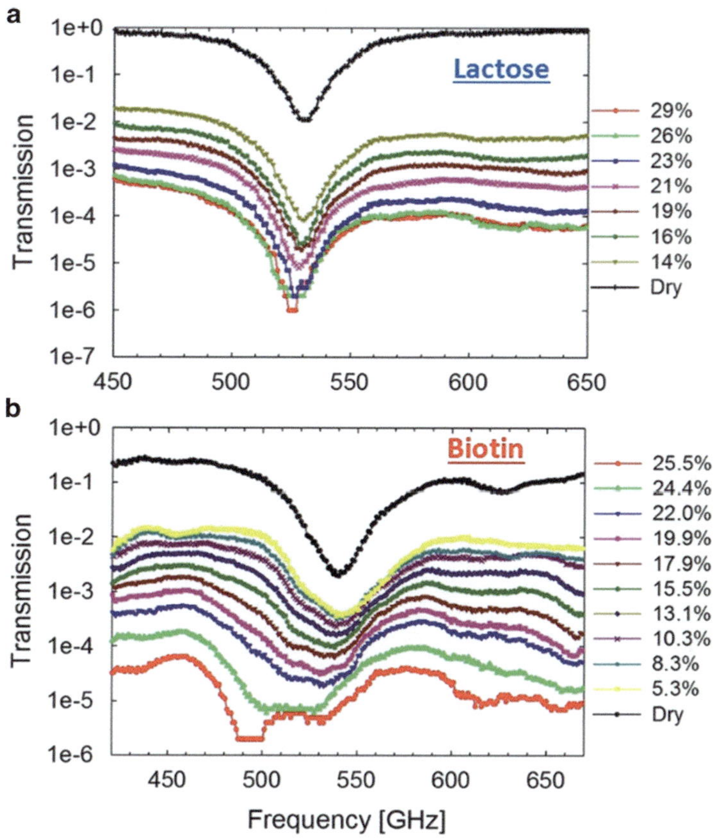

Fig. 13.3 Raw transmission data for two biomolecules chosen for this study: (**a**) lactose monohydrate, and (**b**) biotin

same water lines as above, and the first two are also well-known but weaker water lines. The middle two are similar in shape and strength to the α and β signatures in sample Bt2, but both are red-shifted in center frequency by exactly 11.0 GHz. Given our effective resolution of 1.0 GHz, these cannot be attributed to instrumental drift or frequency error. So we associate them with the same spore-related signatures, α and β, and are labeled as such in Fig. 13.4c. The stationarity of the strong water-vapor lines at 1098 and 1164 GHz support this assertion, and exemplifies the utility of water vapor lines as frequency markers.

Two broader signatures are seen at 510 and 680 GHz. It is tempting to associate these with the 480 and 660 signatures in Fig. 13.4b. However, the shift in frequency of both (30 and 20 GHz, respectively) is much greater than for the α and β signatures, and in the opposite sense (blue shift rather than red shift). They could possibly be associated with vegetative cell material, which is supported by the presence of a broad dip around 480 GHz and the anomalous feature at 700 GHz in

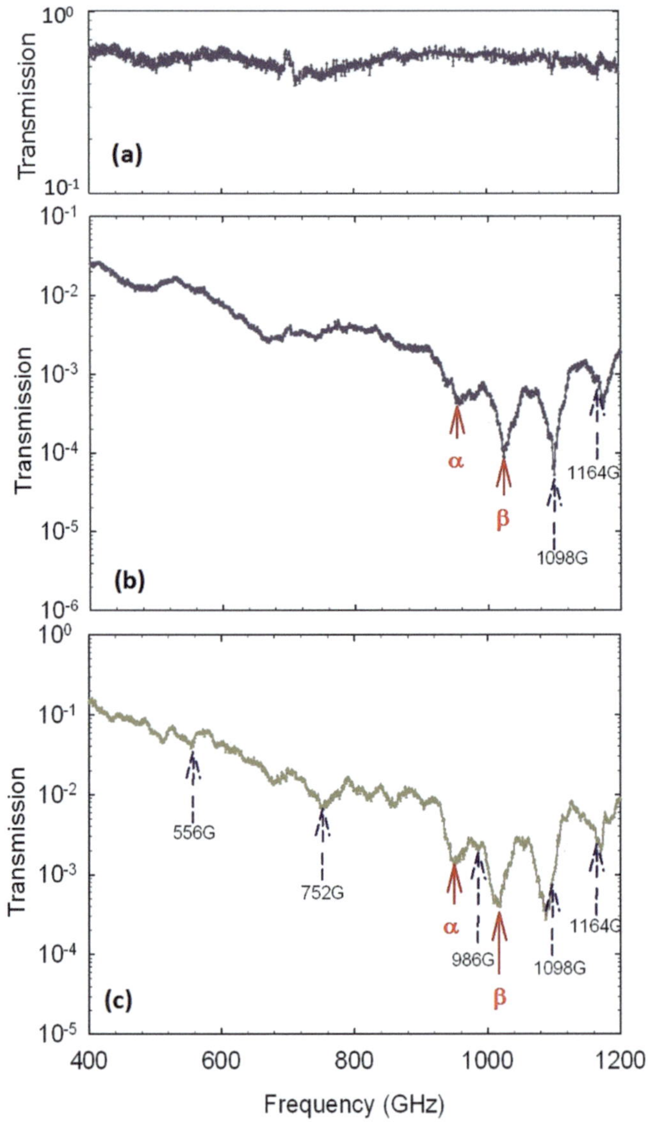

Fig. 13.4 (**a**) Transmission through Bt1 sample at low hydration and with negligible spores. (**b**) Transmission through Bt2 sample at low hydration with spores. (**c**) Transmission through Bt3 sample at moderate hydration with spores

Fig. 13.4a (for the cell-rich, spore-sparse sample Bt1). However, more research is necessary to strengthen this claim.

The transmission spectra of samples Bc1, Bc2, and Bc3 are plotted in Fig. 13.5. Sample Bc1, known to have a significant fraction of spores, shows a predominant signature centered at 880 GHz. It is similar in shape and depth to the β signature of

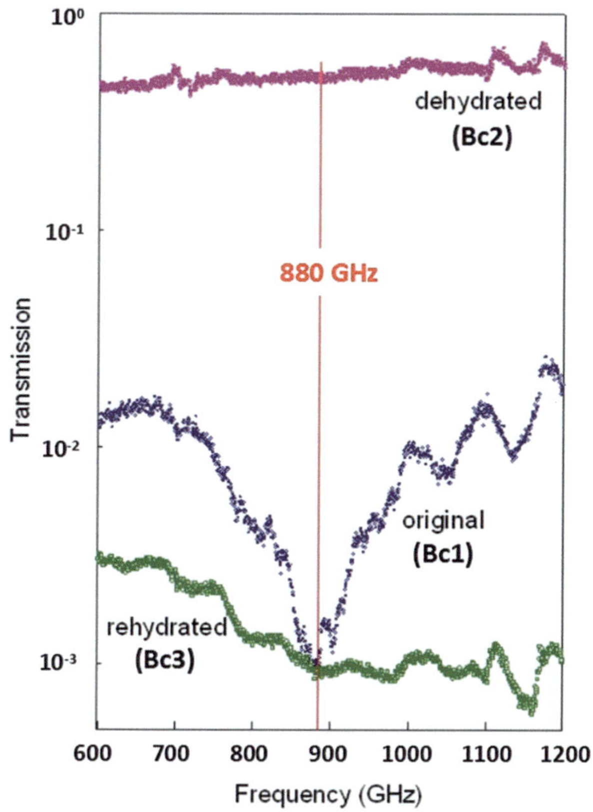

Fig. 13.5 Transmission through Bc sample Bc1 (after preparation at similar hydration as Bt2), Bc2 (dehydrated form of Bc1), and Bc3 (rehydrated form of Bc2)

samples Bt2 and Bt3, but more than 100 GHz lower in frequency. The low background transmission at frequencies well below 880 GHz is consistent with the broadband (Debye) absorption expected from the hydration, and is similar to that for the hydrated Bt samples in Fig. 13.4b, c. When Bc1 was thoroughly dried to become Bc2, the signature completely disappears and the background transmission rises dramatically, similar to sample Bt1 in Fig. 13.4a. Finally, when Bc2 was re-hydrated to levels even greater than Bc1 or Bt3, the resulting Bc3 transmission dropped to the lowest levels seen with Bc2. And the signature seen in Bc1 did not reappear in any form. This suggests that Bc3 had too much hydration, likely much higher than the 25.5% where the biotin signature began to deform in Fig. 13.3b. And it is strong evidence that these Bacillus spore signatures are critically dependent on hydration: with too little hydration, there is no signature like Bc2, and with too much hydration there is no signature like Bc3.

13.4 Analysis

As in most forms of spectroscopy, there is often a wealth of physical information in the lineshape of absorption signatures. Lineshape analysis has already been carried out for the biomolecules [9, 14] and bioparticles [16] studied here under dry or unknown-hydration conditions. In all cases, the line absorption coefficients α were found to be well fit by the famous Lorentzian function for dipole oscillators. The frequency dependence in the most general case is given as

$$\alpha = A \cdot \left[\left(\omega^2 - \omega_0^2 \right)^2 + \omega^2 \gamma^2 \right]^{-1} \tag{13.1}$$

where ω is the circular frequency of incident radiation, ω_0 is the natural circular oscillation frequency (in the absence of damping), γ is the relaxation frequency (i.e., damping constant), and A is a constant that depends on the density and effective charge of the dipoles, and possibly other factors depending on the type of oscillator. Analytic evaluation of the peak absorption frequency ω_r is defined implicitly through $(d\alpha/d\omega)|_{\omega r} = 0$, which leads to

$$\omega_r = \left[(\omega_0)^2 - \gamma^2/2 \right]^{1/2} \tag{13.2}$$

This is monotonically decreasing with respect to γ, meaning that ω_r is always "red-shifted" with respect to the natural frequency. This is a well-known, and intuitive result from classical mechanics and electromagnetics.

Another important prediction of the Lorentzian lineshape function is the linewidth, defined generally as the "fullwidth at half-maximum" (FWHM), or $\Delta\omega$. To find this, we solve for the two ω_0s in (13.1) at which α drops by -3 dB from its peak value [determined by setting $\omega = \omega_r$ in (13.1)], which is $\alpha_p = A[(\omega_0\gamma)^2 - \gamma^4/4]^{-1}$. For $\gamma << \omega_0$, this becomes $\alpha_p \approx A(\omega_0\gamma)^{-2}$, and the -3 dB frequency is found to be $\omega_{-3dB} \approx \omega_0(1 \pm \gamma/2\omega_0)$, so that $\Delta\omega \approx \gamma$.

Given that the Lorentzian model is known to be a good fit to the strongest bio-signatures studied here, we can examine the data with a logical inquiry. The absorption peaks were found to red-shift with increasing hydration, which is also consistent with the Lorentzian model if the damping increases with hydration too. This is a reasonable physical assumption, and if true, then the linewidth should also increase with hydration as $\Delta\omega \approx \gamma$. We will investigate this next for both the biomolecular and bioparticulate signatures.

13.4.1 Biomolecules

Figure 13.6 shows the lineshape results for the experimental lactose-monohydrate and biotin results of Fig. 13.3a, b, respectively. The center frequency in Fig. 13.6a

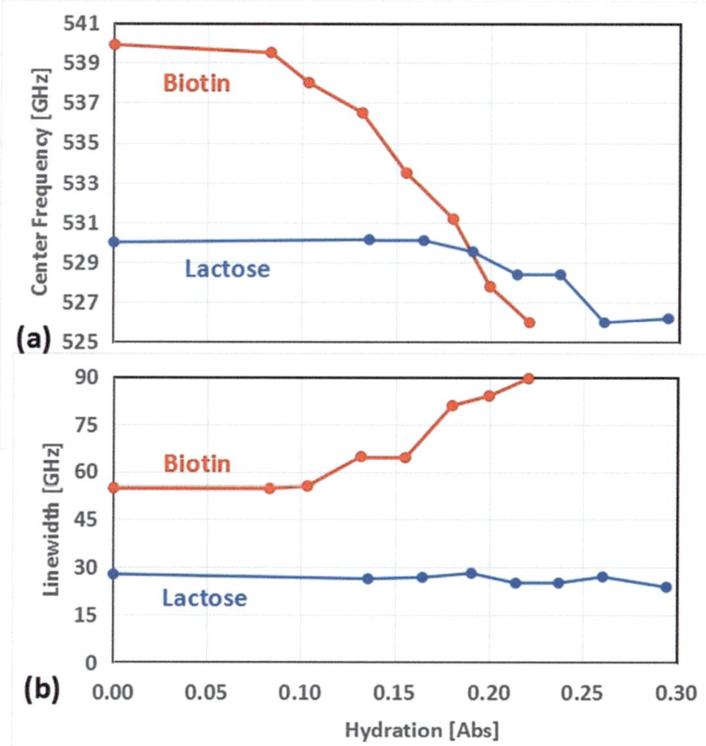

Fig. 13.6 (**a**) Experimental center frequency for the primary resonant signatures of lactose monohydrate (Fig. 13.3a) and biotin (Fig. 13.3b). (**b**) Linewidth (FWHM) calculated by a Lorentzian fit of the signatures in Fig. 13.3a, b

always red-shifts, the biotin frequency dropping rapidly above ~10% hydration, and the lactose not changing significantly until ~20%. The linewidth is obtained by a Lorentzian fitting, and the FWHM (linear frequency) is $\Delta f = \Delta\omega/2\pi = \gamma/2\pi$. The results are plotted in Fig. 13.6b, showing that the biotin at zero hydration is almost twice as broad as the lactose, and then increases dramatically above ~10% consistent qualitatively with the red-shift in center frequency. In contrast, the lactose FWHM does not change significantly, even above the ~20% hydration where the center frequency drops.

To take the analysis one step further, we can use the center frequency from 6 (a) to calculate the Lorentzian linewidth using a rearrangement of Eq. (13.2): $\Delta f = \gamma/2\pi = \{2[(\omega_0)^2 - (\omega_r)^2]\}^{1/2}/2\pi$. This is plotted in Fig. 13.7 in comparison with the experimental Δf for lactose (Fig. 13.7a) and biotin (Fig. 13.7b), using $f_0 = \omega_0/2\pi = 530.4$ GHz (from lactose dry sample), and $f_0 = 541.3$ GHz (biotin dry sample). Clearly, the agreement for lactose is not good, the Lorentz model predicting a significant change in linewidth above ~20% hydration that is not observed experimentally. For biotin, the agreement is better qualitatively, but the

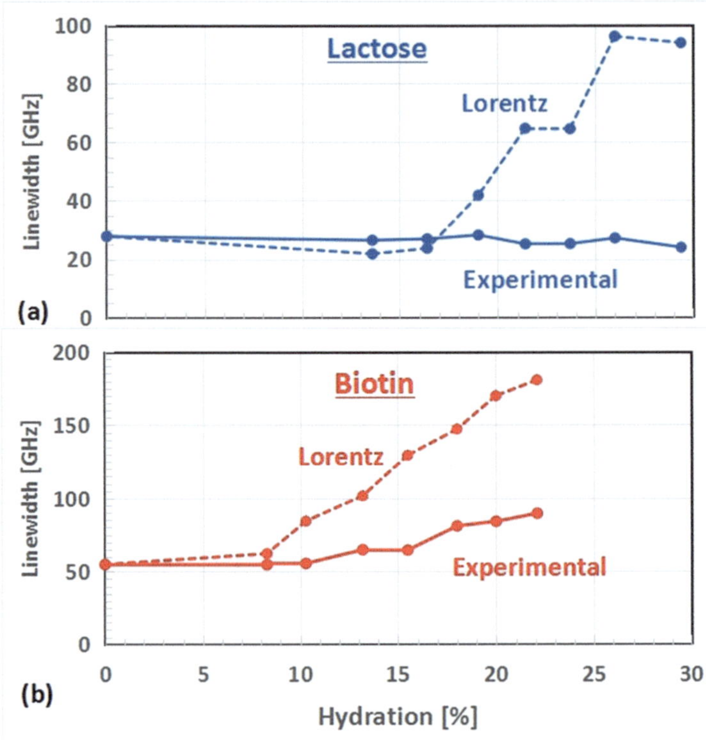

Fig. 13.7 Comparison of experimental linewidth (FWHM) to Lorentzian-fit linewidth for lactose (**a**) and biotin (**b**) samples vs hydration

Lorentz model predicts a linewidth increasing much faster with hydration than the experiment does.

13.4.2 Bioparticles

For the bioparticles (Bacillus endospores), we cannot do such a precise analysis with the Lorentz model because we do not know ω_0, the dry resonant frequency. As mentioned earlier, this is probably because hydration plays two roles with the endospores: (1) induces polarity and (2) increases damping. So no hydration means that no signatures occur in the non-polar endospores. Nevertheless, we can check for consistency with Lorentzian behavior through the experimental center frequency and linewidth for the two signature-displaying samples of Fig. 13.4: Bt2 and Bt3, where the hydration of Bt3 is higher than that of Bt2. Each sample displayed two distinct signatures (labeled α and β in Fig. 13.4) whose center frequencies and FWHM are plotted in Fig 13.8a, b, respectively. Both center

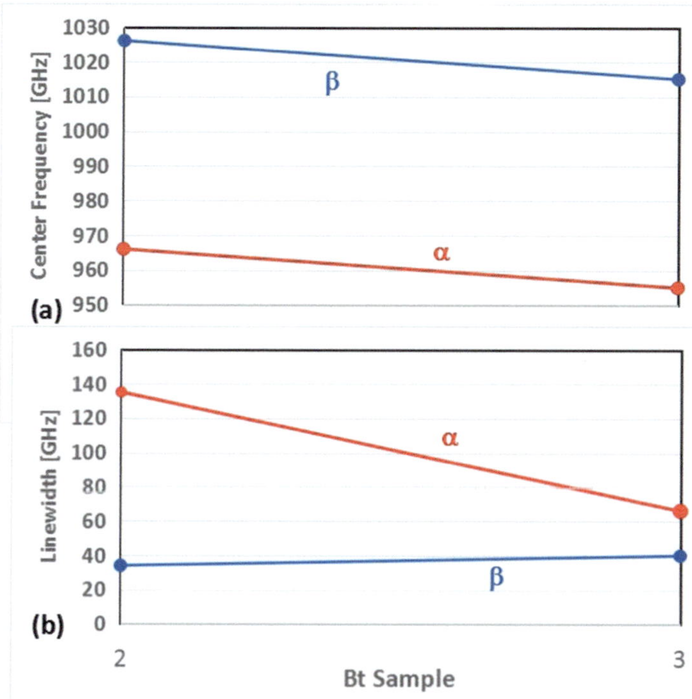

Fig. 13.8 (**a**) Center frequency of signatures α and β in samples Bt2 and Bt3, whose hydration level is known to increase between them. (**b**) Linewidth (FWHM) of signatures α and β vs Bt sample hydration

frequencies decrease by 11 GHz with the increased hydration, consistent with Lorentzian behavior if increased hydration creates increased damping. However, the linewidth behavior in Fig. 13.8b is ambiguous. The β signature linewidth increases with hydration, consistent qualitatively with Lorentzian behavior. But the α signature linewidth decreases with hydration, similar to lactose monohydrate in Fig. 13.6b, but even more opposed. This is something we are investigating further through experiment and modeling.

13.5 Conclusion

We have presented the first known study of the effect of hydration on THz absorption signatures in biomolecules and bioparticles. We observe a "red-shift" in center frequency with increasing hydration in both biomolecules and bioparticles, consistent with Lorentzian-oscillator behavior. But the effect of hydration on linewidth is ambiguous, sometimes increasing the linewidth (consistent with Lorentzian behavior) and sometimes decreasing the linewidth.

Acknowledgement This material is based upon work supported by, or in part by, the U.S. Army Research Laboratory and the U.S. Army Research Office under contract number W911NF-11-1-0024.

References

1. D.C. Lis, J.E. Vaillancourt, P.F. Goldsmith, T.A. Bell, N.Z. Scoville, and J. Zmuidzinas, Submillimeter Astrophysics and Technology, vol. 417 in Astronomical Society of the Pacific Conference Series (Astronomical Society of the Pacific, San Francisco, 2009).
2. J.F. Federici, B. Schulkin, F. Huang, D. Gary, R. Barat, F. Oliveira, and D. Zimdars, THz imaging and sensing for security applications – explosives, weapons, and drugs, Semiconductor Sci. and Tech. 20, S266–S280 (2005).
3. B. Ferguson and X-C. Zhang, Materials for terahertz science and technology, Nature Materials 1, 26–33 (2002).
4. Z.D. Taylor, R.S. Singh, D.B. Bennett, P. Tewari, C.P. Kealey, N. Bajwa, M.O. Culjat, A. Stojadinovic, H. Lee. J-P. Hubschman, E.R. Brown, and W.S. Grundfest, THz Medical Imaging: in-Vivo Hydration Sensing, IEEE Trans. THz Science and Tech. 1, 201–219 (2011).
5. M. Feiginov, C. Sydlo, O. Cojocari, and P. Meissner, Resonant-tunneling-diode oscillators operating at frequencies above 1.1 THz, Appl. Phys. Lett. 99, 235506–235501 (2011).
6. X. Mei, W. Yoshida, M. Lange, J. Lee, J. Zhou, P-H. Liu, K. Leong, A. Zamora, J. Padilla, S. Sarkozy, R. Lai, and W.R. Deal, "First Demonstration of Amplification at 1 THz using 25-nm InP High-Electron Mobility Transistor Process," Elect. Dev. Lett. 36, 327–329 (2015).
7. www.vadiodes.com
8. J.E. Bjarnason and E.R. Brown, "Sensitivity measurement and analysis of an ErAs:GaAs coherent photomixing transceiver," Appl. Phys. Lett. 87, 134105 (2005).
9. A. Roggenbuck, H. Schmitz, A. Deninger, I.C. Mayorga, J. Hemberger, and R. Güsten. Coherent broadband continuous-wave terahertz spectroscopy on solid-state samples, New J. of Physics, 12, 043017 (2010).
10. D. Xu and J.-C. Côté, Sequence Diversity of the Bacillus thuringiensis and B. cereus Sensu Lato Flagellin (H Antigen) Protein: Comparison with H Serotype Diversity, Applied and Environmental Microbiology 72, 4653 (2006).
11. M. Plomp, T. J. Leighton, K. E. Wheeler, and A. J. Malkin, Architecture and high-resolution structure of Bacillus thuringiensis and Bacillus cereus spore coat surfaces, Langmuir, 7892–7898 (2005).
12. H. Sakai, M. T. Howlader, Y. Ishida, A. Nakaguchi, K. Oka, K. Ohbayashi, M. Yamagiwa, and T. Hayakawa, Flexibility and strictness in functional replacement of domain III of cry insecticidal proteins from Bacillus thuringiensis, J. Bioscience and Bioengineering 103, 381 (2007).
13. https://en.wikipedia.org/wiki/Anthrax
14. D.G. Allis, A.M. Fedor, T.M. Korter, J.E. Bjarnason, and E.R. Brown, Assignment of the Lowest-Lying THz Absorption Signatures in Biotin and Lactose Monohydrate by Solid-State Density Functional Theory, Chem Phys. Lett, 440, 203–209 (2007).
15. M. van Exter, C. Fattinger, and D. Grischkowsky, Terahertz time-domain spectroscopy of water vapor. Opt. Lett. 14, 1128–1130 (1989).
16. W-D. Zhang, E.R. Brown, L. Viveros, K.P. Burris, and C. Neal Stewart, Jr., Narrow terahertz attenuation signatures in Bacillus thuringiensis, J. Biophotonics 1–7 (2013).

Chapter 14
Anisotropic Medium Approach for the Optical Nonlinearities of Dilute Nitride Superlattices

C.I. Oriaku, T.J. Spencer, and M.F. Pereira

Abstract In this paper we investigate optical nonlinearities in semiconductor superlatitces designed with alternate layers of dilute nitride, GaAsN quantum wells and AlGaAs and barriers. The anisotropic medium approach is used to obtain analytical expressions. The analysis confirms that the nonlinearities increase with anisotropy, extending the anisotropic medium approach to dilute nitride systems.

Keywords Many body effects • Optical nonlinearities • Superlattices • Anisotropic medium approach

14.1 Introduction

Semiconductor superlattices are ideal structures to control and study transport [1] and optical properties of semiconductor structures of varying effective dimensionality. This paper addresses the investigation of optical nonlinearities in dilute nitride semiconductor superlattices by means of the anisotropic medium approach, which can be easily programmed and includes the main many body effects required to describe steady state nonlinear absorption [2]. This has potential for a major impact in the development of new materials with applications from terahertz (THz) and mid-infrared (MIR) to the visible (VIS) spectral range [3]. The superlattices are modelled as anisotropic media characterized by effective masses parallel and perpendicular to the growth direction [2]. That suggests that new materials being currently investigated for solar cells and which are strongly anisotropic, may also

C.I. Oriaku • T.J. Spencer • M.F. Pereira (✉)
Materials and Engineering Research Institute, Sheffield Hallam University, S1 1WB Sheffield, UK

Department of Physics, Michael Okpara University of Agriculture, Umudike, P.M.B. 7267, Umuahia, Nigeria
e-mail: M.Pereira@shu.ac.uk

M.F. Pereira, O. Shulika (eds.), *THz for CBRN and Explosives Detection and Diagnosis*, NATO Science for Peace and Security Series B: Physics and Biophysics,
DOI 10.1007/978-94-024-1093-8_14

be very useful for nonlinear optics applications. Here the anisotropy-induced nonlinearity in the nanostructure is controlled per design in contrast with less controllable recent studies in which it depends on the nanoparticle shape [4]. We show that the nonlinearities increase with the anisotropic character, but the results are only valid as long as sufficient tunnelling occurs. In other words, isolated quantum wells should not be addressed by this method. This paper extends and complements the study in Refs. [5, 6] with applications to a different type of material and is organized as follows: first, the main equations used are summarized. Next, numerical applications for $GaAs_xN_{1-x}/Al_yGa_{1-y}As$ superlattices are given before a short summary.

14.2 Mathematical Approach and Model Equations

The carrier-induced optical nonlinearities of semiconductors can be obtained by the self-consistent evaluation of Many Body Nonequilibrium Green's Functions (NEGF). Efficient numerical methods used here have been successfully applied to both inter-subband [7–12] and inter-band transitions [13–16] in quantum wells and superlattices. For completeness, we repeat here the main Eqs. of Ref. [6] for a guide to the reader, by starting with the interband polarization that has been used to describe superlattices as an effective anisotropic 3D material [2],

$$\left[\hbar\omega + i\Gamma - E_g(\gamma) - \frac{\hbar^2 k^2}{2\mu_{\|}}\xi(\theta,\gamma)\right] P\left(\vec{k},\omega\right) A(\omega) \left[\wp \cdot E(\omega) \sum_{\vec{k}' \neq \vec{k}} \mathcal{W}_{\vec{k}-\vec{k}'} P\left(\vec{k}',\omega\right)\right], \tag{14.1}$$

where the meaning of the anisotropy dependent bandgap is explained below and $\xi(\theta,\gamma) = \sin^2\theta + \gamma\cos^2\theta$. The anisotropy parameter γ is given by the ratio between the in-plane $\mu_{\|}$ and perpendicular $\mu_{\perp}$ reduced effective masses, $\gamma = \mu_{\|}/\mu_{\perp}$, with $\frac{1}{\mu_{\|}} = \frac{1}{m_{e\|}} + \frac{1}{m_{h\|}}$ and $\frac{1}{\mu_{\perp}} = \frac{1}{m_{e\perp}} + \frac{1}{m_{h\perp}}$ which are calculated from the non-interacting superlattice Hamiltonian $\mathcal{H}_0$, $\frac{1}{m_{i\|}} = \hbar^{-2}\partial^2/\partial k_{i\|}^2 \langle\Psi|\mathcal{H}_0|\Psi\rangle$, $\frac{1}{m_{i\perp}} = \hbar^{-2}\partial^2/\partial k_{i\perp}^2 \langle\Psi|\mathcal{H}_0|\Psi\rangle$, for i = e,h. More details are given in Ref. [2]. A phenomenological scattering rate Γ is used to simulate the average dephasing that stems from the electron-electron, electron-phonon and electron-impurity scattering [17–19], in order to keep the approach as simple as possible without affecting the conclusions. We have neglected any k-dependence on the transition dipole moment $\wp$ induced by the electric field E(ω). In a superlattices there is a preferred direction determined by the growth (z-direction), leading to cylindrical symmetry. The next step is to perform an angle average, $\langle\xi(\theta,\gamma)\rangle = \langle\sin^2\theta + \gamma\cos^2\theta\rangle = \frac{1}{2}(1+\gamma)$ and thus $\frac{\hbar^2 k^2}{2\mu_{\|}}\xi(\theta,\gamma) \longrightarrow \frac{\hbar^2 k^2}{2M_{\|}}$.

The resulting anisotropic mass $M_{||} = 2\mu_{||}/(1+\gamma)$determines the exciton Bohr radius and corresponding 1S binding energy, $a_0 = a_0(\gamma) = \hbar^2\epsilon_0/e^2M_{||}$, $E_0 = E_0(\gamma) = 2Ry/(1+\gamma)$, $Ry = \frac{\mu_{||}\,e^4}{2\hbar^2\epsilon_0^2}$, where ϵ_0 and e denote, respectively the background dielectric constant and the electron charge. The resulting exciton binding energies in superlattices are in very good agreement with experiments, as demonstrated in Ref. [2]. By approximating the full inversion factor by $A(\omega) = \tanh[\beta(\hbar\omega - \mu)/2]$ we make sure that the cross-over from absorption to gain takes place exactly at the total chemical potential. Next, we replace the usual Yukawa potential describing screened carriers in 3D (which does not have analytical solutions for the corresponding Schrödinger equation) by the Hulthén potential $\mathcal{W}(r) = -2e^2\kappa\epsilon_0'^{-1}/((\exp(2\kappa r) - 1))$ [2, 5, 6, 20, 21], which has successfully reproduced bulk nonlinear optical spectra [18] and has well known analytical solutions. The total chemical potential $\mu = \mu_e + \mu_h$ is given by $\beta\mu_\lambda = \ln v_\lambda + K_1 \ln(K_2 v_\lambda + 1) + K_3\mu_\lambda$. The screening wavelength is $= \kappa_e + \kappa_h$, where for each particle $\kappa_\lambda = \sqrt{\frac{4\pi e^2}{\epsilon_0}\frac{\partial n_\lambda}{\partial \mu_\lambda}} = \sqrt{\frac{\beta\pi e^2 n_\lambda/\epsilon_0}{\frac{1}{v_\lambda} + \frac{K_1 K_2}{K_2 v_\lambda + 1} + K_3}}$, where $K_1 = 4.897$, $K_2 = 0.045$ and $K_3 = 0.133$. The parameter v_λ is obtained from the particle density for electrons and holes, $n = n_e = n_h$ by $v_\lambda = 4n_\lambda/[(2m_{||,\lambda}/\beta\pi\hbar^2)(2m_{\perp,\lambda}/\beta\pi\hbar^2)^{1/2}]$ and $\beta = 1/K_BT$ [2].

It is beyond the scope of this paper to show the intermediate details that lead Eq. (14.2). The main steps are to combine the partly phenomenological approach of Ref. [21] with the material parameters calculated with the anisotropic medium approach [2, 22]. Furthermore, we use hypergeometric functions [21] instead of the Jost function approach of Refs. [2, 20]. The resulting absorption spectrum then reads

$$\alpha(\omega) = \alpha_0(\gamma)\frac{\hbar\omega}{e_0}\tanh[\beta(\hbar\omega - \mu)/2]\left\{\sum_{n=1}^{\sqrt{g}}\frac{4\pi}{n}\left(\frac{1}{n^2} - \frac{n^2}{g^2}\right)\delta_\Gamma(\Delta - e_n)\right.$$
$$\left. +2\pi\int_0^\infty \frac{\sinh(\pi g\sqrt{x})}{\cosh(\pi g\sqrt{x}) - \cos\left(\sqrt{4g - g^2x}\right)}\delta_\Gamma(\Delta(-x))dx\right\}, \quad (14.2)$$

where the prefactor $\alpha_0(\gamma)$ and normalised detuning Δ are given by $\alpha_0(\gamma) = 2\wp^2/\left(n_b\hbar c a_0^3(\gamma)\right)$, $\Delta = (\hbar\omega - E_g(\gamma))/E_0(\gamma)$. The background refractive index and the speed of light in vacuum are given respectively by n_b and c. In Eq. (6), the band gap renormalization is given by an approximation that mimics the results obtained with the static Single Plasmon Pole Approximation (SPPA) [2].

$$E_g(\gamma) = E_{gap} + E_0(\gamma)\begin{cases} -1 + \left(1 - \frac{1}{g(\gamma)}\right)^2, & g(\gamma) \geq 1 \\ -1/g(\gamma) & , g(\gamma) < 1 \end{cases} \quad (14.3)$$

where $E_{gap} = E_c(1) + E_{HH}(1) + E_{gap,\,bulk}$ is the bulk (temperature dependent) isotropic bandgap including the confinement energies of the lowest conduction subband and the top heavy hole subband and where $g(\gamma) = 1/\kappa a_0(\gamma)$.

We chose $\delta_\Gamma(x) = 1/\pi\delta\eta\cosh(x/\eta)$, where $\eta = \Gamma/E_0(\gamma)$ to represent the broadened Dirac delta function. The sum in the exciton part runs through the available states up to the largest integer value of $\sqrt{g(\gamma)}$.

14.3 Results and Discussion

The material selected for the superlattices in this numerical superlattice is made of dilute nitride quantum wells and AlGAs barriers, specifically $GaAs_{0.99}N_{0.01}$/ $Al_{0.3}Ga_{0.7}As$ alternating layers. The dilute nitride composition is so small that we assume that the layers are unstrained as in the usual GaAs case and thus a superlattice can be actually grown. Ref. [11] has details of bandstructure calculations for the QW case and references on how to obtain the bulk Hamiltonian giving rise to the effective masses used here as input to the anisotropic medium approach [2, 5, 6]. The increase in anisotropy in the anisotropic material is obtained by increasing the barrier width. Table 14.1 summarizes the relevant set of parameters used in the numerical results shown next

Figure 14.1 depicts the evolution of the anisotropic parameter as a function of barrier width if the well with is held constant. The progressively stronger localization of wavefunctions increases the z-direction effective masses and thus reduces the anisotropic parameter γ.

Figure 14.2 shows the nonlinear optical absorption for different dilute nitride superlattices superlattices at $T = 300$ K. Larger anisotropy (smaller γ) obtained by increasing the quantum well length, stems from reducing the wavefunction overlap through the barrier. The single quantum well case would be obtained for a very large barrier that yields $\gamma \to 0$, but this is still an effective 3 medium and thus, this method should not be used in the quasi-2D case. For truly isolated quantum wells, the 2D or quasi-2D limit should be used, see e.g. Ref. [14]). But the important outcome is that from (a) to (e) we do see an increase in nonlinearity, i.e. a larger reduction in absorption with increasing carrier density.

Figure 14.3 shows differential absorption, $\Delta\alpha(\omega) = \alpha(\omega, N) - \alpha(\omega, N=0)$.

Table 14.1 $GaAs_xN_{1-x}$/ $Al_yGa_{1-y}As$ anisotropic superlattices ($x = 1\%$ and $y = 30\%$)

L_b (nm)	L_w (nm)	γ	Eg0 (eV)
0.1	10	0.683	1.2482
1	10	0.460	1.2685
2.5	10	0.154	1.2755
4	10	0.044	1.2768
7.5	10	0.002	1.2770
10	10	2.44×10^{-4}	1.2771

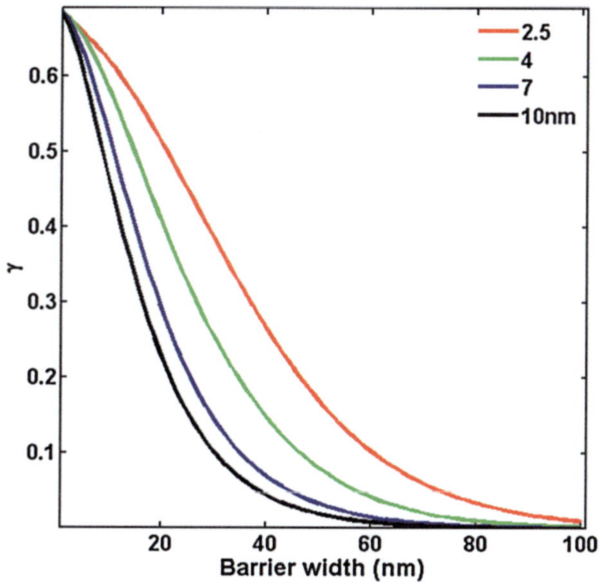

Fig. 14.1 Anisotropy parameter for $(GaAs)_xN_{1-x}As/Al_yGa_{1-y}As$ superlattices as a function of barrier width. The well widths considered are: (a) 2.5, (b) 4, (c) 7 and (d) 10 (nm) respectively. The mole fractions x and y are 1% and 30% respectively were used in the calculation

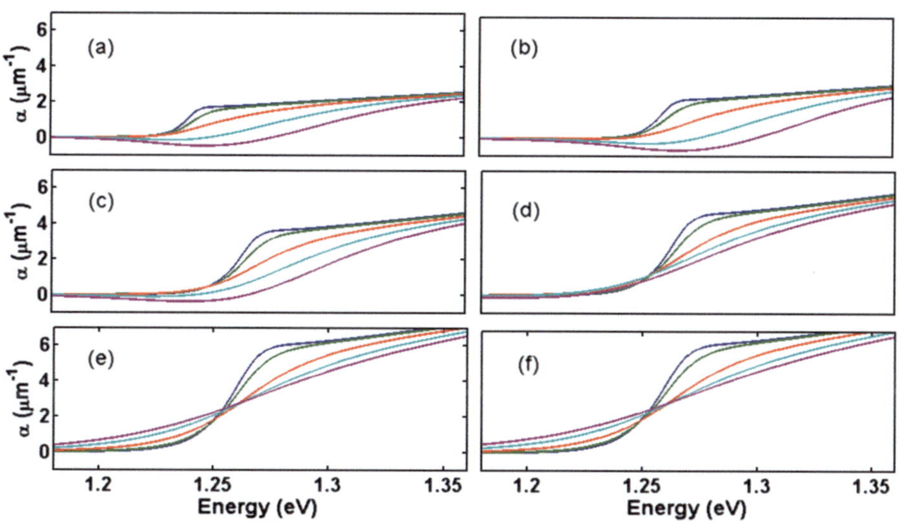

Fig. 14.2 Nonlinear Optical absorption of $(GaAs)_xN_{1-x}As/Al_yGa_{1-y}As$ superlattices at 300 K for $\gamma = 0.683, 0.46, 0.154, 0.044, 0.002, 2.44 \times 10^{-4}$, respectively from (**a**) to (**f**). In each panel, from *top to bottom* the carrier density in both conduction and valence bands is N = 0, 0.1, 0.5, 1 and 1.5×10^{18} cm^{-3}

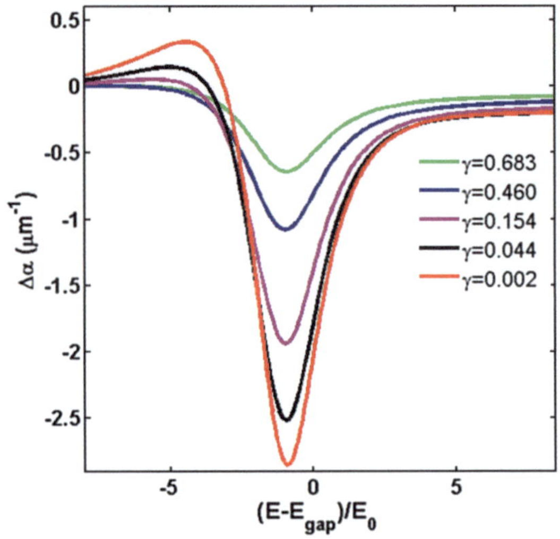

Fig. 14.3 Differential absorption $\Delta\alpha(\omega) = \alpha(\omega, N) - \alpha(\omega, N = 0)$ for the structures in Fig. 14.1 with a carrier density $N = 1 \times 10^{17}\ cm^{-3}$ at 300 K

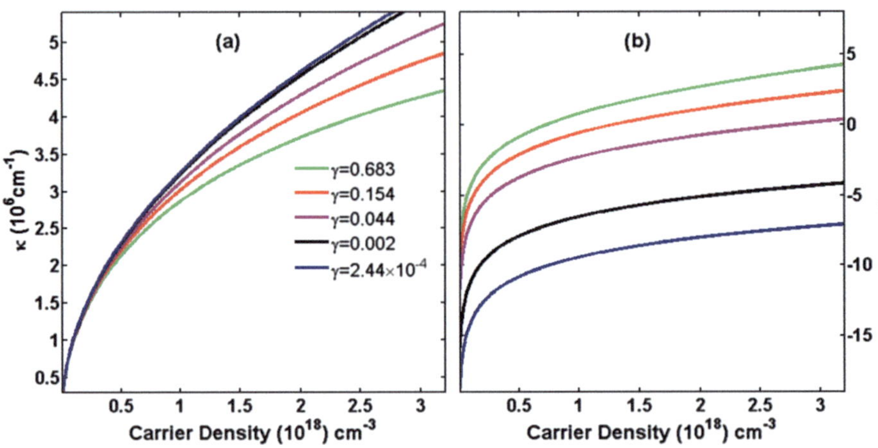

Fig. 14.4 Plots of room temperature (**a**) screening wavenumber and (**b**) quasi-chemical potential versus the carrier density of $GaAs_xN_{1-x}/Al_yGa_{1-y}As$ superlattices for the different anisotropy parameters

Further understanding of the excitonic bleaching process is delivered by Fig. 14.4, which depicts the screening length potential for different structures as a function of carrier density.

The ensemble of results in the figures above show that larger anisotropy (smaller γ), increases the inverse screening length leading to a faster reduction of the screened Coulomb interaction and consequently larger optical nonlinearity, measured here directly by the differential absorption.

Note however that it only makes sense to consider the range of linear to saturated absorption in the differential transmission calculations while two positive values are compared. The negative absorption (gain) confuses the comparison.

By increasing the anisotropy, the z-direction electron and hole masses become too different, with a much larger increase in hole mass. When the upper and lower (average) curvatures or equivalently, the average effective masses are too different the total chemical potential is relatively smaller (see Fig. 14.4) and so is the inversion factor $A(\omega) = \tanh[\beta(\hbar\omega - \mu)/2]$ of Eq. (14.1), thus reducing the gain in Fig. 14.1. This influence of different electron and hole masses on the inversion/gain is entirely consistent with the detailed analysis for isolated quantum wells seen e.g. in Ref. [16].

14.4 Conclusion

In summary, the analytical expressions for optical nonlinearities in effective anisotropic media were extended to dilute nitride superlattices. This study further confirms the connection between an increase in optical nonlinearity with anisotropy by directly controlling the anisotropy and evaluating the resulting differential absorption and nonlinear spectra. This illustrates the potential for three dimensional strongly anisotropic semiconductors materials for a role in nonlinear optics and a possible recipe to improve their efficiency by controlling the anisotropy.

Acknowledgements The authors acknowledge support from COST ACTION MP1204, TERA-MIR Radiation: Materials, Generation, Detection and Applications.

References

1. A. Wacker, Semiconductor superlattices: a model system for nonlinear transport, Physics Reports 357, 1–111 (2002).
2. M.F. Pereira Jr., Analytical solutions for the optical absorption of superlattices, Phys. Rev. B 52, 1978–1983 (1995).
3. M.F. Pereira, TERA-MIR radiation: materials, generation, detection and applications II, Opt Quant Electron 47, 815–820 (2015).
4. Y. Hua, K. Chandra , D.H. Dam, G.P. Wiederrecht and T.W. Odom, Shape-dependent nonlinear optical properties of anisotropic gold nanoparticles, J Phys Chem Lett. 6, 4904–4908 (2015).
5. M. F. Pereira, Anisotropy and Nonlinearity in Superlattices, Optical and Quantum Electronics 48, 321-1–321-7 (2016).
6. M. F. Pereira, Anisotropy and Nonlinearity in Superlattices II, Optical and Quantum Electronics 48, 423-1 – 423 - 8(2016).
7. M.F. Pereira and I.A. Faragai, Coupling of THz radiation with intervalence band transitions in microcavities, Optics Express 22, 3439–3446 (2014).
8. M. F. Pereira Jr. Intersubband antipolaritons: microscopic approach, Phys. Rev. B 75, 195301-1–195301-5 (2007).

9. M. F. Pereira Jr., Intervalence electric mode terahertz lasing without population inversion, Phys. Rev. B 78, 245305-1–245305-5 (2008).
10. M.F. Pereira Jr, R. Nelander, A. Wacker, D.G. Revin, M.R.Soulby, L.R. Wilson, J.W. Cockburn, A.B. Krysa, J.S. Roberts, and R.J. Airey, Characterization of intersubband devices combining a nonequilibrium many body theory with transmission spectroscopy experiments, Journal of Materials Science: Materials in Electronics 18, 689–694 (2007).
11. M. F. Pereira and S. Tomić, Intersubband gain without global inversion through dilute nitride band engineering, Appl. Phys. Lett. 98, 061101-1–061101-3 (2011).
12. M.F. Pereira Jr., Microscopic approach for intersubband-based thermophotovoltaic structures in the THz and Mid Infrared, JOSA B28, 2014–2017 (2011).
13. H. Grempel, A. Diessel, W. Ebeling, J. Gutowski, K. Schuell, B. Jobst, M.F. Pereira Jr., and K. Henneberger, High-density effects, stimulated emission and electrooptical properties of ZnCdSe/ZnSe single quantum wells and laser diodes, Phys. Stat. Sol. B194, 199–217 (1996).
14. M.F. Pereira Jr., R. Binder and S.W. Koch, Theory of nonlinear absorption in coupled band quantum wells with many-body effects, Appl. Phys. Lett. 64, 279–281 (1994).
15. M.F. Pereira Jr. and K. Henneberger, Gain mechanisms and lasing in II–VI compounds, Phys. Stat. Sol. B202, 751–762 (1997).
16. W.W. Chow, M.F. Pereira Jr., and S.W. Koch, Many-body treatment on the modulation response in a strained quantum well semiconductor laser medium, Appl. Phys. Lett. 61, 758–760 (1992).
17. T. Schmielau and M.F. Pereira Jr., Nonequilibrium many body theory for quantum transport in terahertz quantum cascade lasers, Appl. Phys. Lett. 95, 231111-1–231111-3 (2009).
18. T. Schmielau and M.F. Pereira, Impact of momentum dependent matrix elements on scattering effects in quantum cascade lasers, Physica Status Solidi B 246, 329–331 (2009).
19. T. Schmielau and M.F. Pereira, Microelectronics Journal 40, 869–871 (2009).
20. L. Bányai and S.W. Koch, A Simple theory for the effects of plasma screening on the optical spectra of highly excited semiconductors, Z. Physik B63, 283–291 (1986).
21. C.I. Oriaku and M.F. Pereira, Analytical solutions for semiconductor luminescence including Coulomb correlations with applications to dilute bismides, J. Opt. Soc. Am B 34, 321–328 (2017).
22. M.F. Pereira Jr., I.Galbraith, S.W.Koch, and G.Duggan, Exciton binding energies in semiconductor superlattices: an anisotropic effective-medium approach, Phys. Rev. B42, 7084–7089 (1990).

Part III
THz Sources

Chapter 15
Development of Terahertz Frequency Quantum Cascade Lasers for the Applications as Local Oscillators

Y.J. Han, L.H. Li, A. Valavanis, N. Brewster, J.X. Zhu, R. Dong, P. Dean, L. Bushnell, M. Oldfield, A.G. Davies, B. Ellison, and E.H. Linfield

Abstract We report the development of terahertz frequency quantum cascade lasers for applications as local oscillators. A range of active region designs and waveguide structures have been characterised in order to develop the devices for operation at high temperatures, with high output power and low dissipated power. Quantum cascade lasers based on a LO-phonon bound-to-continuum design emitting at 3.5 THz, suitable for the detection of hydroxyl, were fabricated with a double-metal (gold-gold) waveguide structure. These devices operated in continuous-wave up to 94 K, with an output power of 0.4 mW and dissipated power of 1.7 W at 10 K. A new, mechanically robust packaging and waveguide-integration scheme is also presented for operation outside laboratory environments, which further allows integration of quantum cascade lasers with terahertz waveguides, mixers and other system components. This integration scheme yielded a better beam quality, with a divergence of <20°, compared to standard double-metal devices. Its impacts on the device performance, such as operating temperature range, spectral emission, output power and electrical properties, are presented.

Keywords Terahertz • Quantum cascade laser • Local oscillator

15.1 Introduction

A range of chemical, biological, radiological and nuclear defence (CBRN) agents and explosives exhibit vibrational and rotational transitions at THz and mid-infrared (TERA-MIR) frequencies, and this has fuelled interest in developing

---

Y.J. Han (✉) • L.H. Li • A. Valavanis • J.X. Zhu • R. Dong • P. Dean • A.G. Davies
E.H. Linfield
School of Electronic and Electrical Engineering, University of Leeds, Leeds LS2 9JT, UK
e-mail: y.han@leeds.ac.uk

N. Brewster • L. Bushnell • M. Oldfield • B. Ellison
Rutherford Appleton Laboratory, STFC, Harwell Oxford, Didcot OX11 0QX, UK

M.F. Pereira, O. Shulika (eds.), *THz for CBRN and Explosives Detection and Diagnosis*, NATO Science for Peace and Security Series B: Physics and Biophysics,
DOI 10.1007/978-94-024-1093-8_15

materials, devices and systems for generating and detecting TERA-MIR radiation [1, 2]. Quantum-cascade lasers [3] (QCLs), in particular, are promising light sources in the terahertz (THz) spectral region [4]. After more than a decade's development, THz QCLs now cover the frequency range of 1–5 THz [5–7], and have been demonstrated as local oscillators in heterodyne receivers [8, 9]. Indeed, high resolution heterodyne spectroscopy is a powerful tool to measure molecular rotational lines and fine structure lines of atoms or ions. Many of these lines are located at the THz frequency region, including the hydroxyl (OH) rotational transitions at 1.8 THz, 2.5 THz, and 3.5 THz, and the atomic oxygen (OI) fine structure line at 4.7 THz, which are useful for exploring the Earth's atmosphere and interstellar media [10, 11]. Our aim is to use THz QCLs as local oscillators in airborne and satellite heterodyne spectrometers. For practical reasons, cooling of THz QCLs by liquid helium should ideally be replaced by a mechanical closed-cycle cooler. To be suitable as local oscillators in this configuration, THz QCLs have to simultaneously meet a number of essential requirements, such as maintaining a precise and targeted lasing frequency, high operating temperature, high optical output power, low dissipated power, good stability and a good beam profile. At the same time, a mechanically robust and reproducible scheme for integrating THz QCLs with external waveguides and mixers is required.

Here, we present the fabrication of THz frequency quantum cascade lasers (QCLs) emitting at 3.5 THz, as well as a new and robust integration scheme. The paper is organized as follows. In Sect. 15.2, we discuss the details of the materials growth and device processing. In Sect 15.3, we present a comparison of THz QCLs with different active region designs and different waveguide structures, and the characterisation results of a high-performance 3.5 THz laser. In Sect. 15.4, we introduce a new and robust integration scheme for THz QCLs [12], and present its effects on the beam profile, electrical and optical output properties. Finally, we discuss some further optimization of QCLs and integrations, and draw some conclusions.

15.2 Device Fabrication and Characterisation

The QCLs used in this work are based on GaAs/AlGaAs materials. The QCL samples were grown on semi-insulating GaAs substrates using molecular beam epitaxy. For each wafer, a GaAs buffer layer and a 300 nm-thick $Al_{0.5}Ga_{0.5}As$ etch-stop layer were followed by a 700 nm-thick n^+-GaAs contact layer doped with Si to 2×10^{18} cm^{-3}. After those layers, the active region structure was grown repeatedly to a total thickness of 10–15 μm. On the top of the active region, a 50 nm-thick n^+-GaAs contact layer was grown with a Si concentration of 5×10^{18} cm^{-3}.

QCL devices were processed according to the configuration of a semi-insulating-substrate-surface-plasmon (SISP) waveguide [4] or a double-metal (DM) waveguide [13]. For the SISP waveguide, laser ridges were defined by wet chemical etching. A 150 nm-thick AuGeNi electrical bottom contact layer was deposited beside the ridges and annealed at 400 °C for 1 min. Two 10 μm-wide stripes of AuGeNi layers were deposited on the ridges as a top contact. The waveguide metal layers, Ti/Au (10/150 nm), were deposited on top of the laser ridges and annealed at 270 °C for 4 min. For the DM waveguide, Ti/Au of thickness 10/500 nm was deposited on top of the QCL wafer, and also on a second n+-GaAs acceptor wafer. The two wafers were then thermal compression bonded at the metallic interfaces. The substrate of the QCL wafer was lapped to 10 μm, and removed by selective wet chemical etching. Following wet-etching of the laser ridges, Ti/Au (10/150 nm) was deposited on the top. The substrate was thinned and metallised to improve thermal performance, and the ridges were then cleaved and Indium-soldered on the copper submounts.

For the light intensity–current–voltage (L–I–V) and spectral characterization, the devices were mounted on the cold finger of a helium-cooled cryostat and operated in continuous wave (CW) mode or in pulsed mode with 2-μs-long pulses at a repetition rate of 10 kHz. The laser emission was detected using a cooled Ge:Ga bolometer. The output power was calibrated using an absolute THz power meter (Thomas–Keating Instruments).

15.3 Experimental Results of QCLs

In airborne or satellite heterodyne spectrometers, the dissipated power of THz QCLs should be below the cooling power of the closed-cycle cooler, which is typically several watts at 80 K and decreases at lower temperatures. A QCL with low dissipated power, high operating temperature and high output power is therefore desirable for the optimisation of spectrometer systems. Starting with a comparison of different active region designs and waveguide structures, we here identify a suitable design and present characterisation data for a high-performance device.

15.3.1 Comparison of Active Region Designs

Comparison of different active region designs is not straight-forward since those designs normally have different lasing frequencies, doping concentration and active region thickness. Material growth and fabrication processes also affect the device performance. In order to make the comparison easier, three commonly investigated designs lasing around 3.3 ± 0.2 THz were grown and fabricated under similar conditions. The differences in doping concentration or total thickness can be implied by the device dissipated power and the output power.

In Fig. 15.1, the performance of QCL devices based on a bound-to-continuum (B-to-C), B-to-C with longitudinal-optical (LO) phonon extraction (hybrid) and resonant phonon (RP) designs are compared [14–16]. All three devices were

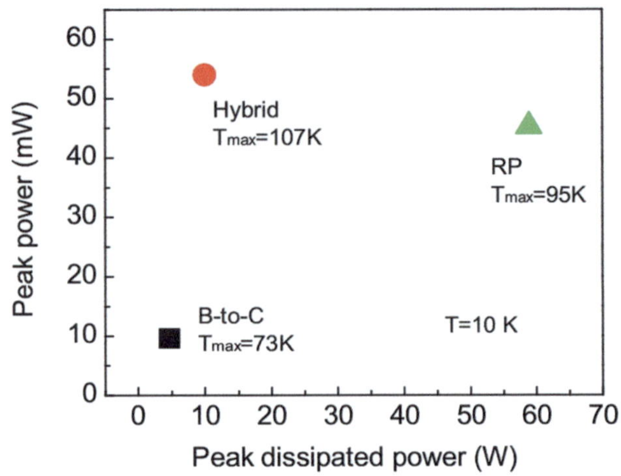

Fig. 15.1 Comparison of three active region designs. The devices have the same dimensions of 3000 μm × 150 μm, and were operated in pulsed mode

processed into the SISP waveguide, cleaved to the same size of 3000 μm × 150 μm, and characterised in pulsed mode. The B-to-C design had the lowest dissipated power (4.6 W), which satisfies the requirement of the provided cooling power. However, it gave the lowest maximum operating temperature (73 K) and the lowest output power (9.7 mW), limiting its use as a local oscillator, especially at high temperatures. The RP design yielded a higher maximum operating temperature (95 K) and a higher output power (45.1 mW), but the dissipated power was much higher than the B-to-C design. In this case, the increased electrical power consumption generated much more Joule heating inside the RP structure, and limited the CW performance. The required cooling power for the RP design was about 60 W, which is out of the capability of typical closed-cycle coolers. Although the dissipated power can be lowered by decreasing the doping density or the device size, it is at the expenses of the output power. The hybrid design was found to provide the benefits of both the B-to-C design and the RP design; it yielded the highest output power (54 mW) and the highest maximum operating temperature (107 K) of the three structures, and at the same, the dissipated power was comparable to the B-to-C design. Therefore, the hybrid design is identified as a promising choice to meet local oscillator requirements.

15.3.2 Comparison of Waveguide Structures

Two waveguide configurations, SISP and DM, are normally used for THz QCLs [4, 13]. In the SISP waveguide, a semi-insulating substrate and a thin confinement layer lowers the waveguide loss at the expense of the confinement factor (and hence net gain) of guided modes. In the DM waveguide, the optical field is almost entirely confined inside the active region by two metal layers, but this strict confinement results in widely divergent emission.

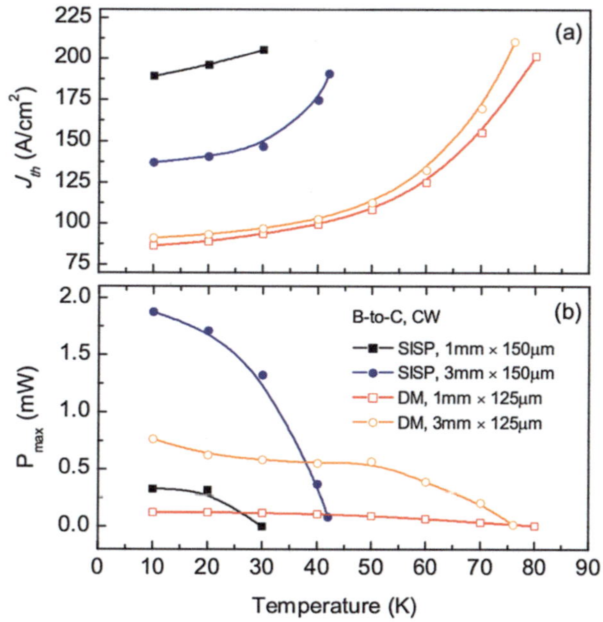

Fig. 15.2 Temperature evolution of (**a**) threshold current density, J_{th}, and (**b**) the maximum output power, P_{max}, of THz QCLs with different waveguide configuration. The devices were operated in CW mode

Figure 15.2 shows the temperature dependence of the CW threshold current density, J_{th}, and maximum output power of two pairs of laser ridges process with a SISP waveguide and a DM waveguide, respectively. The only difference between the devices in each pair is the length, being 1 mm for one device and 3 mm for the other. The width of devices in each pair is the same, 150 μm for the SISP devices and 125 μm for the DM devices. As can be seen, the effects of employing double-metal waveguide are significant. Increases of more than 30 K in the maximum operating temperature were observed, with the output power only degrading significantly for temperatures higher than 50 K. Although the output power of the DM devices was only at half the level obtained with the SISP devices at 10 K, the slow thermal damping rate of DM devices resulted in higher optical power at elevated temperatures. With the increase of device length, the mirror loss is lowered. As a result, a reduction in threshold current density was observed for both waveguide structures, and the decrease of the threshold current is more obvious for the SISP device due to the smaller optical mode confinement factor. The longer devices emitted higher optical output power, but also dissipated more electrical power. This caused increased Joule heating, limiting the maximum operating temperature of the long DM QCL to 4 K lower than that of the short device. Overall, the DM QCLs showed better temperature performance than the SISP QCLs, i.e., higher output power at high temperatures, and therefore, the DM waveguide is preferred to be employed to develop QCLs for local oscillator applications.

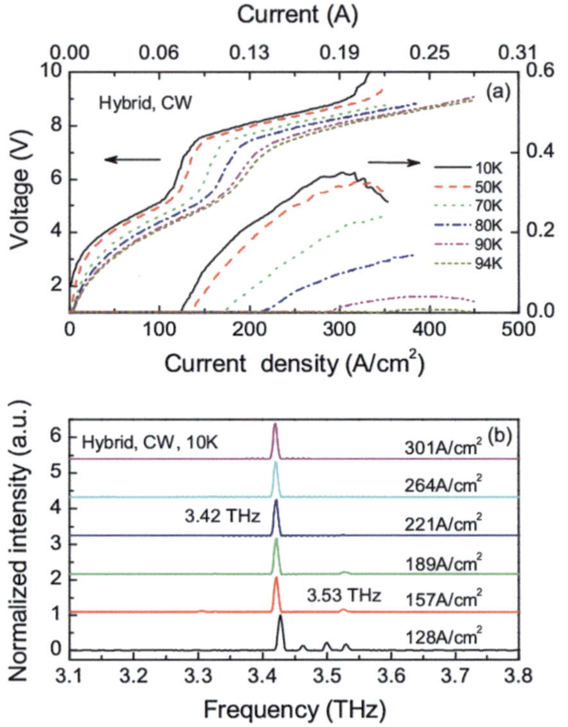

Fig. 15.3 L–I–V curves at different temperatures and emission spectra at 10 K. The device was operated in CW mode

15.3.3 QCLs Lasing at 3.5 THz

The hybrid active region plus the DM waveguide have been identified as the optimum choice for THz QCLs for use as local oscillators, owing to the good temperature performance and low dissipated power. Based on the hybrid design shown in Fig. 15.1, the active region was modified to shift the gain spectra from 3.1 THz to 3.5 THz. During the material growth, the total thickness was also increased to 15 μm and the doping concentration in the active region was optimized. The device was processed into DM waveguide structure with dimensions of 500 μm × 125 μm, and driven in CW mode.

Figure 15.3 shows L–I–V characteristics obtained at different temperatures, as well as the emission spectra at 10 K. This optimised design yielded a considerable improvement in the operating temperature and the output power. Compared with the B-to-C design (Fig. 15.2), the maximum operating temperature increased from 80 K to 94 K, and the output power from 0.13 mW to 0.35 mW. The dissipated power was 1.7 W, higher than the B-to-C design, but within the requirement of cooling power. Furthermore, this device provided more than 0.1 mW output power with a dissipated power of 2.0 W at 80 K. Note that the output power presented here is an underestimate of the generated THz power, since the collection efficiency is not considered. The emission spectra was observed from 3.42 to 3.53 THz, and covered the target frequency of 3.5 THz.

15.4 Integration of QCLs with Waveguide Blocks

QCLs in double-metal waveguides yield good thermal performance, as demonstrated above, but also poor far-field beam quality and wide beam divergence. Although the far-field beam can be improved by using techniques such as device patterning approaches [17, 18] or assemblies of antennas or lenses [19, 20], their reproducibility and mechanical robustness have not been demonstrated. Here, we present a waveguide block integration scheme for THz QCLs [12], and investigate its effects on the beam profile and laser performance.

15.4.1 Waveguide Integration Scheme

In the waveguide-integration scheme [12], a double-metal QCL was mounted within a copper heat-sink enclosure containing a rectangular cross-section metallic waveguide. The QCL was based on the B-to-C design, as described in Fig. 15.2, and the device length was adjusted to 980 μm. The QCL chip contained 13 independent ridges although only one of them was connected electrically.

Figure 15.4 shows the waveguide block with a QCL integrated inside. The waveguide block was precision-machined directly into a Cu block by mechanical microfabrication techniques, with a channel of cross-sectional dimensions 300 μm $\times$ 150 μm, including a cavity for the positioning of a QCL. The QCL chip was mounted in the cavity by using an indium foil to ensure a good thermal contact. The output facet of the central QCL ridge was aligned with the waveguide aperture. A Au ribbon bond was used to provide an electrical connection to the top contact of the QCL through a small Au-plated strip of AlN and a Au-plated quartz substrate secured to the Cu block by high conductivity silver-loaded epoxy. The complete assembled block was mounted into a cryostat, and the electrical bias was supplied through a subminiature-A (SMA) connector on the rear of the block. The temperature was monitored at the top of the block.

To measure the far-field beam profile of the integrated QCL, a Golay detector with an entrance aperture of 1 mm diameter was scanned linearly in two dimensions at a distance 32.4 mm from the laser facet. The laser was driven using a continuous 30 Hz square wave, and operated at a heat-sink temperature of 30 K. As shown in Fig. 15.5, an approximately Gaussian single-lobed profile was observed. The full-width at half-maximum (FWHM) in the growth and in-plane directions is 20° and 17°, respectively [12]. This is comparable to the divergence obtained previously by patterning the QCL or through the use of device-integrated antenna structures or lenses, and the 'ringing' that is commonly associated with the beam profiles of normally mounted DM QCLs is removed entirely.

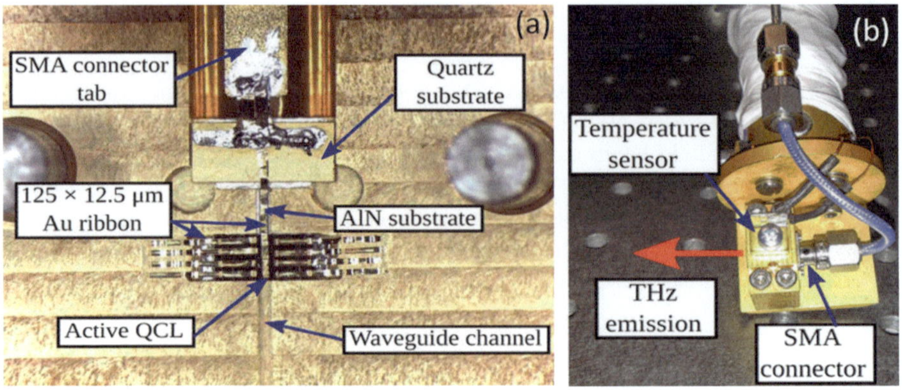

Fig. 15.4 (**a**) Interior of bottom half of waveguide block, and (**b**) exterior of QCL waveguide block (Reproduced from Valavanis et al. [12])

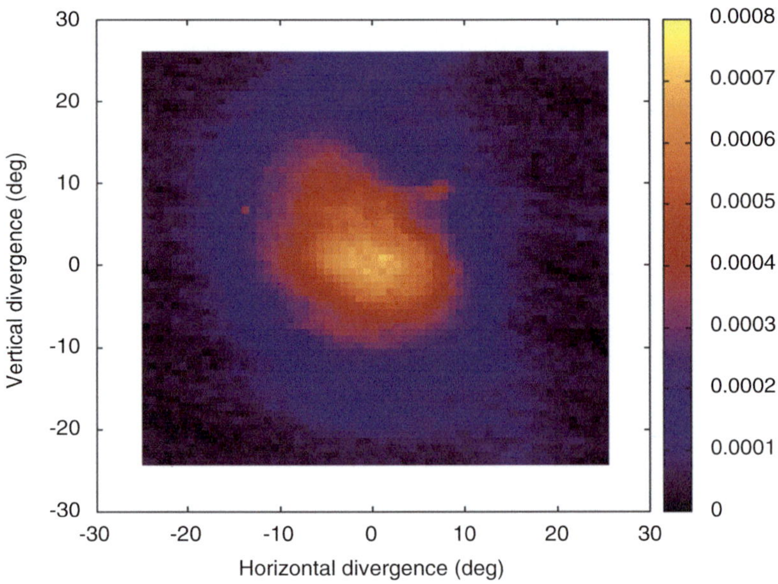

Fig. 15.5 Two dimensional image of far-field beam profile (Reproduced from Valavanis et al. [12])

15.4.2 Comparison with Standard Mounted QCLs

The temperature performance and the spectra of the integrated QCL have been compared with those of a standard mounted QCL, in order to investigate the effects of the waveguide block upon the maximum operating temperature, output power, dissipated power and electrical connections. The standard mounted QCL used for this comparison is the short DM device reported in Fig. 15.2, with a length of 1 mm being the only difference to the QCL integrated in the waveguide block (length 980 μm).

Figure 15.6 shows the L–I–V curves of the standard QCL and the integrated QCL at different temperatures. In comparison, the I–V curves of the two devices are similar, expect for a drop in the applied voltage of the integrated QCL. At a temperature of 20 K, the voltage drop, ΔV, can be fitted to the function $\Delta V = I \times R_s + V_s$, from which a series resistance $R_s = 1.2\ \Omega$ and an addition voltage drop $V_s = 0.9$ V are obtained. This implies that the integration scheme improved the electrical connections and will lead to a reduction in dissipated power. The integrated QCL yielded a threshold current density of 102 A/cm^2 at 20 K and maximum operation temperature of 77 K, slightly worse compared with 89 A/cm^2 and 80 K of the standard device. This can be attributed to a small increase in the thermal resistance of the waveguide block and cryostat. The integrated device was also found to have a reduction in the output power to 32 μW, about 25% of the standard device. Sincc thcrc was no significant change in I–V curves, the drop in the output power, has been ascribed to a mismatch between the transverse mode profile of the laser cavity and that of the relatively wide rectangular waveguide, and also between the waveguide aperture and free space [12].

Figure 15.7 shows the emission spectra of the standard and integrated QCLs. Multimode emission was observed in both cases, with the same central frequency of 3.4 THz and similar spectral bandwidth of ~120 GHz. The side modes emitted from the integrated device were broadened and attenuated, which could be related with the waveguide channel in the block, rather than the QCL itself.

Overall it can be seen that this integration scheme can improve the emission beam quality significantly, as well as the electrical connection, while not introducing significant perturbation to the internal properties of the laser itself. The issues related to the reduction of the output power arising from the mismatch between the transverse mode profile of the laser cavity and that of the waveguide, will be addressed in the next design.

15.5 Conclusion

We have reported the development and optimisation of terahertz frequency quantum cascade lasers for applications as local oscillators. A device consisting of a 'hybrid' active region design based on a bound-to-continuum transition with longitudinal-optical (LO) phonon extraction, plus a double metal waveguide structure, has been identified as a promising choice. A high-performance QCL based on this design and emitting at 3.5 THz was fabricated, and shown to operate in CW mode up to a temperature of 94 K, with a low temperature output power of ~0.4 mW and dissipated power of 1.7 W. Single frequency emission and better thermal management will be considered in the next device iteration. In addition, a new waveguide integration scheme for THz QCLs was demonstrated based on a highly

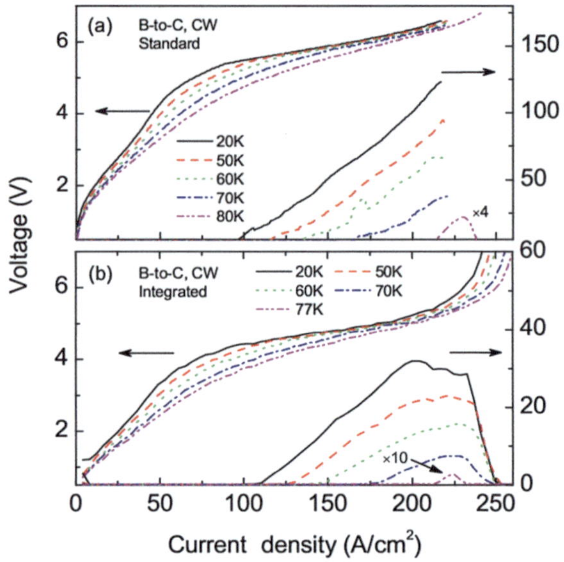

Fig. 15.6 L–I–V curves of (**a**) standard mounted and (**b**) integrated QCLs at different temperatures. The lasers operated in CW mode

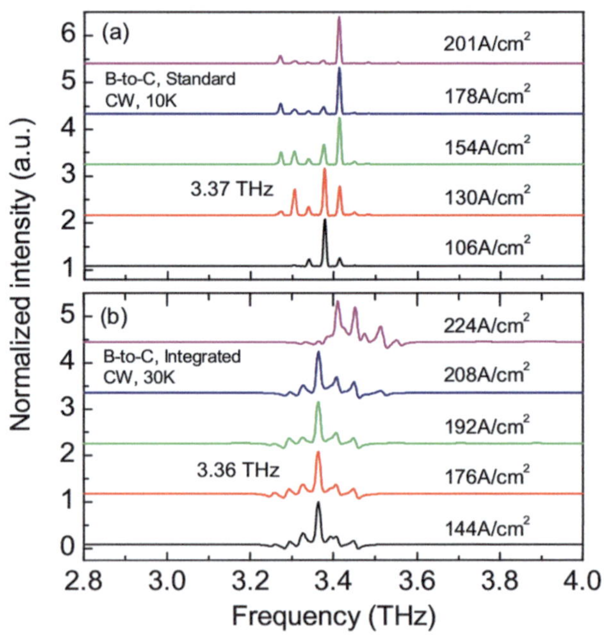

Fig. 15.7 Emission spectra of (**a**) standard mounted and (**b**) integrated QCLs. The lasers were operated in CW mode

reproducible precision-micromachining technique. This approach improved the electrical connections and yielded a considerably improved beam divergence of $<20^\circ$, whilst maintaining a low threshold current and good operating temperature range. A higher output power can be expected by integrating an optimized QCL with a new design of waveguide block.

Acknowledgments We acknowledge financial support from the EPSRC (UK) 'COTS' programme, the ERC grant 'NOTES' and 'TOSCA', NERC (UK), the European Space Agency, and the CEOI-ST. AGD and EHL also acknowledge support from the Royal Society and the Wolfson Foundation. PD acknowledges support from EPSRC (UK).

References

1. http://www.tera-mir.org
2. P. Dean, A. Valavanis, J. Keeley, K. Bertling, Y. L. Lim, R. Alhathlool, A. D. Burnett, L. H. Li, S. P. Khanna, D. Indjin, T. Taimre, A. D. Rakić, E. H. Linfield, A. G. Davies, Terahertz imaging using quantum cascade lasers – review of systems and applications, J. Phys. D 47, 374008 (2014)
3. J. Faist, F. Capasso, D. L. Sivco, C. Sirtor, A. L. Hutchinson, A. Y. Cho, Quantum Cascade Laser, Science 264, 553–556 (1994)
4. R. Köhler, A. Tredicucci, F. Beltram, H. E. Beere, E. H. Linfield, A. G. Davies, D. A. Ritchie, R. C. Iotti, F. Rossi, Terahertz semiconductor-heterostructure laser, Nature 417, 156–159 (2002)
5. L. Li, L. Chen, J. Zhu, J. Freeman, P. Dean, A. Valavanis, A. G. Davies E.H. Linfield, Terahertz quantum cascade lasers with >1 W output powers, Electron Lett. 50, 309–311 (2014)
6. C. Walther, M. Fischer, G. Scalari, R. Terazzi, N. Hoyler, J. Faist, Quantum cascade lasers operating from 1.2 to 1.6 THz, Appl. Phys. Lett. 91, 131122 (2007)
7. C. W. Chan, Q. Hu, J. L. Reno, Ground state terahertz quantum cascade lasers, Appl. Phys. Lett. 101, 151108 (2012)
8. J. R. Gao, J. N. Hovenier, Z. Q. Yang, J. J. A. Baselmans, A. Baryshev, M. Hajenius, T. M. Klapwijk, A. J. L. Adam, T. O. Klaassen, B. S. Williams, S. Kumar, Q. Hu, J. L. Reno, Terahertz heterodyne receiver based on a quantum cascade laser and a superconducting bolometer, Appl. Phys. Lett. 86, 244104 (2005)
9. H. W. Hubers, S. G. Pavlov, A. D. Semenov, R. Köhler, L. Mahler, A. Tredicucci, H. E. Beere D. A. Ritchie, E. H. Linfield, Terahertz quantum cascade laser as local oscillator in a heterodyne receiver, Opt Express 13, 5890–5896 (2005)
10. R. G. Prinn, J. Huang, R. F. Weiss, D. M. Cunnold, P.J. Fraser, P.G. Simmonds, A. McCulloch, C. Harth, P. Salameh, S. O'Doherty, R.H.J. Wang, L. Porter, B. R. Miller, Evidence for substantial variations of atmospheric hydroxyl radicals in the past two decades, Science 292, 1882–1888, (2001)
11. R. T. Boreiko1, A. L. Betz1, Heterodyne Spectroscopy of the 63 μm O I Line in M42, The Astrophysical J. Lett. 464, L83 (1996)
12. A. Valavanis, Y. J. Han, N. Brewster, P. Dean, R. Dong, L. Bushnell, M. Oldfield, J. X. Zhu, L. H. Li, A. G. Davies, B. Ellison, E. H. Linfield, Mechanically robust waveguide-integration and beam shaping of terahertz quantum cascade lasers, Electron. Lett. 51, 919–920 (2015)
13. B. S. Williams, S. Kumar, H. Callebaut, Qing Hu, J. L. Reno, Terahertz quantum-cascade laser at lambda approximate to 100 mu m using metal waveguide for mode confinement, Appl. Phys. Lett. 83, 2124–2126 (2003)
14. G. Scalari, L. Ajili, J. Faist, H. Beere, E. H. Linfield, David Ritchie, A. G. Davies, Far-infrared (lambda similar or equal to 87 μm) bound-to-continuum quantum-cascade lasers operating up to 90 K, Appl. Phys. Lett. 82, 3165–3167 (2003)

15. M. Wienold, L. Schrottke, M. Giehler, R. Hey, W. Anders, H. T. Grahn, Low-voltage terahertz quantum-cascade lasers based on LO-phonon-assisted interminiband transitions, Electron. Lett. 45, 1030–1040 (2009)
16. S. Fathololoumi, E. Dupont, C. W. Chan, Z. R. Wasilewski, S. R. Laframboise, D. Ban, A. Mátyás, C. Jirauschek, Q. Hu, and H. C. Liu, Terahertz quantum cascade lasers operating up to ~ 200 K with optimized oscillator strength and improved injection tunneling, Opt. Express 20, 3866–3876 (2012)
17. M. I. Amanti, M. Fischer, G. Scalari, M. Beck, J. Faist, Low-divergence single-mode terahertz quantum cascade laser, Nature Photonics 3, 586–590 (2009)
18. N. F. Yu, J. Fan, and Q. J. Wang, C. Pflügl, L. Diehl, T. Edamura, M. Yamanishi, H. Kan, F. Capasso, Small-divergence semiconductor lasers by plasmonic collimation, Nature Photonics 2, 564–570 (2008)
19. M. I. Amanti, M. Fischer, C. Walther, G. Scalari, M. Beck, J. Faist, Horn antennas for terahertz quantum cascade lasers, Electron Lett 43, 573–574 (2007)
20. A. W. M. Lee, Q. Qin, S. Kumar, B. S. Williams, Q. Hu, J. L. Reno, High-power and high-temperature THz quantum-cascade lasers based on lens-coupled metal-metal waveguides, Opt. Lett. 32, 2840–2842 (2007)

Chapter 16
Broadband Heterogeneous Quantum Cascade Lasers

Manijeh Razeghi and Neelanjan Bandyopadhyay

Abstract Heterogeneous quantum cascade lasers have emerged as an important broadband radiation source for mid-infrared spectroscopy and sensing. Multiple quantum cascade stages with discrete peak wavelengths are grown together in single active region to have a broad gain. Different wavelength laser can be chosen by appropriate feedback mechanism. Heterogeneous strain-balanced $Al_{0.63}In_{0.37}As$/$Ga_{0.35}In_{0.65}As$/$Ga_{0.47}In_{0.53}As$ quantum cascade lasers, having 5 and 6 quantum cascade stages respectively, has been demonstrated with gain between 6.2–9.9 μm and 5.9–10.9 μm in mid-infrared region.

Keywords Quantum cascade lasers • Broadband laser • Semiconductor lasers • Mid-infrared

16.1 Introduction

Quantum cascade laser(QCL) [1] are in high demand due to applications in explosive detection, pollution control and industrial process monitoring [2, 3]. Since its invention, QCLs have made rapid progress in terms of spectral coverage [4–6] and output power at room temperature (RT). This has in turn led to enormous progress in the field of mid-infrared spectroscopy.

In recent years, heterogeneous quantum cascade lasers(HQCL) [7] has become popular due to wide mid-infrared spectral emission and the possibility of monolithic integration of a spectrometer on a single chip. In comparison to single core QCL, a HQCL demonstrates much wider emission wavelengths. By wavelength selective feedback like distributed feedback grating [8] or external cavity [7] grating the laser wavelength can be tuned over a wide wavelength. The HQCL has led to the possibility of ultra-wideband spectrometer which can be used to

M. Razeghi (✉) • N. Bandyopadhyay
Center for Quantum Devices, Department of Electrical Engineering and Computer Science, Northwestern University, Evanston, IL 60208, USA
e-mail: razeghi@eecs.northwestern.edu

M.F. Pereira, O. Shulika (eds.), *THz for CBRN and Explosives Detection and Diagnosis*, NATO Science for Peace and Security Series B: Physics and Biophysics,
DOI 10.1007/978-94-024-1093-8_16

identify multiple molecules having their peak absorption features at separate wavelengths.

16.2 Materials and Methods

16.2.1 HQCL

A heterogeneous QCL is a vertically integrated QCL structure with multiple QC stages designed to lase at different wavelengths. Each QC stage covers a small portion of the emission band. Altogether, the entire structure can cover an extremely broad wavelength range. All laser cores share the same waveguide. The confinement factor is subject to design change by changing the number of stages of each core. The minimum number of cores is determined by the gain bandwidth of each QC stage. For the typical bandwidth of mid-IR QCL, a minimum of 5 QC stages are necessary to achieve a reasonably flat gain spectrum.

16.2.2 QCLs with Same Material Compositions

Previously, the active regions were designed in lattice matched $Al_{0.48}In_{0.52}As$/$Ga_{0.47}In_{0.53}As$ so that a very thick broadband structure comprising of many cores can be grown without strain relaxation. While this was successful for longer wavelength emission, the performance of lattice matched QCLs deteriorates at shorter wavelength due to insufficient electron confinement. Strain-balanced structures can solve this problem, but the strain levels, hence material compositions, are usually changing with emission wavelength due to the difference in well and barrier widths. Changing the composition dynamically during a single growth run in a standard molecular beam epitaxy (MBE) reactor presents a very complex structure, which is difficult to characterize with standard techniques [5]. The large number of unknowns (individual layer growth rates and composition) leads to uncertainty and poor quality control. This would lead to uncertainty in the emission wavelengths of individual cores, and make it extremely difficult to produce a reproducibly flat gain spectrum.

An alternate design scheme, in which the material composition of all the cores in a broadband QCL is kept constant, is therefore desirable, provided high efficiency can be maintained at all wavelengths. To address this issue, a new design approach was put forward where the compositions of highly strained $Al_{0.63}In0.{}_{37As}$/$Ga_{0.35}In_{0.65}As$ were kept constant, while the net strain in the structure can be tuned by using composite wells of $Ga_{0.35}In_{0.65}As/Ga_{0.47}In_{0.53}As$. With this basis, (Fig. 16.1) the electron confinement is sufficient for emission above $\lambda = 5$ μm, and calibration is straightforward.

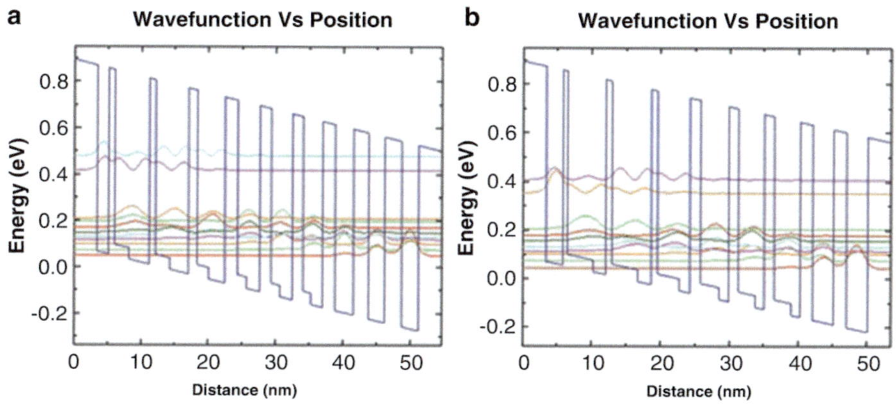

Fig. 16.1 Conduction band diagrams of composite well lasers designed for emission at (**a**) $\lambda = 6.7$ μm and (**b**) $\lambda = 9.1$ μm

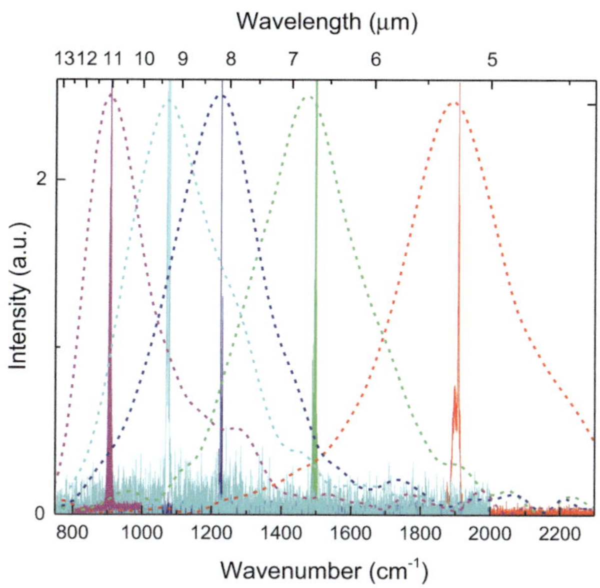

Fig. 16.2 Electroluminescence and laser emission spectrum of the composite well QCLs

By adjusting layer thicknesses only, QCs stages were designed with emission wavelengths from 5.2 to 11 μm. The maximum wall plug efficiencies are 18.4%, 15.7%, 15.4%, 13.2%, 4.8% for 5.2, 6.7, 8.2, 9.1, 11 μm, respectively. These values are comparable to the state-of-the-art. The EL and emission spectrum of the lasers just beyond threshold are shown in the Fig. 16.2.

16.2.3 Waveguide Modelling

The tunable laser waveguide requires both an equalized gain curve and a grating host layer for the sampled grating. Due to the sparse grating coverage, though, the coupling strength must be significantly higher than for a traditional DFB QCL. A coupling coefficient on the order of 100 cm^{-1} provides sufficient design flexibility. Assuming a square grating profile, the coupling coefficient can be estimated by the following formula:

$$\kappa = \frac{2\Delta n_{eff}}{\lambda_0} \tag{16.1}$$

where Δn_{eff} is the difference in effective refractive index of the laser waveguide with and without the grating layer, λ_0 is the free space wavelength. Using a 1-D dispersive model, this quantity can be rapidly calculated as a function of wavelength for a variety of different core and grating layer thicknesses.

As the multi-QC stage gain is less intense that a single core, it is necessary to increase the total number of emitting stages to keep the laser threshold low. A HQCL, as drawn in Fig. 16.3, uses a 3.4 μm thick core to have a high optical confinement factor. As a result, a thicker grating layer is also required to achieve sufficient grating coupling. A 750 nm thick grating layer is reasonable for tunable laser.

16.2.4 Gain Modelling

The net gain curve $G(\hbar\omega)$ of the HQCL is calculated by the sum of individual QC stage gain $g_k(\hbar\omega)$ of the k th stage, weighted by the respective modal confinement factors $\Gamma_k(\hbar\omega)$ at different $\hbar\omega$.

$$G(\hbar\omega) = \sum_{k=1}^{N} \Gamma_k(\hbar\omega) g_k(\hbar\omega) \tag{16.2}$$

The gain of the kth stage is given by relation

$$g_k(\hbar\omega) = \sum_{i,j} g_{ji}(\hbar\omega) \tag{16.3}$$

where the gain $g_{ji}(\hbar\omega)$ due to transition between levels j, i is calculated from the first order gain model by the expression,

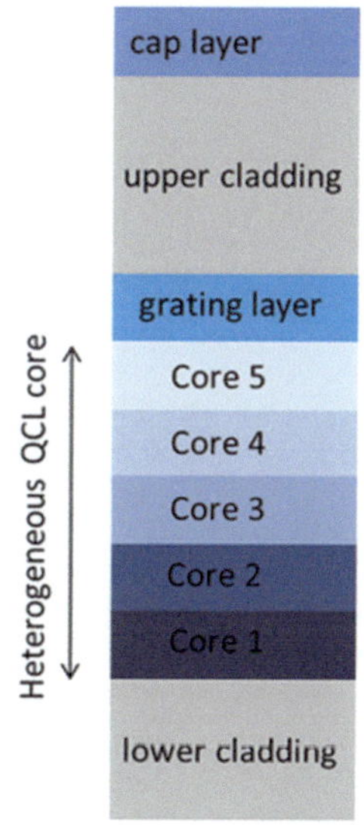

Fig. 16.3 Schematic of 5 QC stage HQCL

$$g_{ji}(\hbar\omega) = \frac{\omega_{ji}^2}{\omega}\left(\frac{e^2}{n_r c\varepsilon_0}\right)\frac{{z_{ji}}^2\left(\Gamma_{ji}/_2\right)}{\left(\hbar\omega_{ji}-\hbar\omega\right)^2+\left(\Gamma_{ji}/_2\right)^2}\left(N_j - N_i\right) \quad (16.4)$$

where $\hbar\omega_i$ and N_i represent energy and electron concentration of ith level respectively; $\hbar\omega$ is the energy of the electromagnetic wave $g_{ji}(\hbar\omega)$ is the energy dependent gain due to transition of dipole strength z_{ji} between two levels j and i, separated by energy $\hbar\omega_{ji}$ and broadened by linewidth Γ_{ji} .e,c, ε_0, n_r are the electronic charge, speed of light, vacuum permittivity and refractive index respectively.

In order to design the HQCL, we proceeded step by step by incorporating multiple stages starting from 5 stages. HQCL consisting of 5 & 6 QC stages with peak gain at emission energies spaced ~20 meV apart, are used to form the HQCL core. For 6 stage HQCL, the QC stages, starting from longest wavelength stage, had a total of 8, 9, 9, 10, 10, 10 wells, respectively, with designed resonant fields of 61, 61, 67, 67, 72, 78 kV/cm, respectively. The 5 stage QCL did not have the longest wavelength stage compared to 6 stage HQCL.

The gain of the HQCL at resonant field is demonstrated in Fig. 16.4 for (a) 5 and (b) 6 QC stage HQCL. By appropriate choice of active QC stages and number of periods of each stage a flat gain was obtained over the wavelength of operation.

The EL of the 5 and 6 QC HQCL is displayed in Fig. 16.5 (a) and (b).

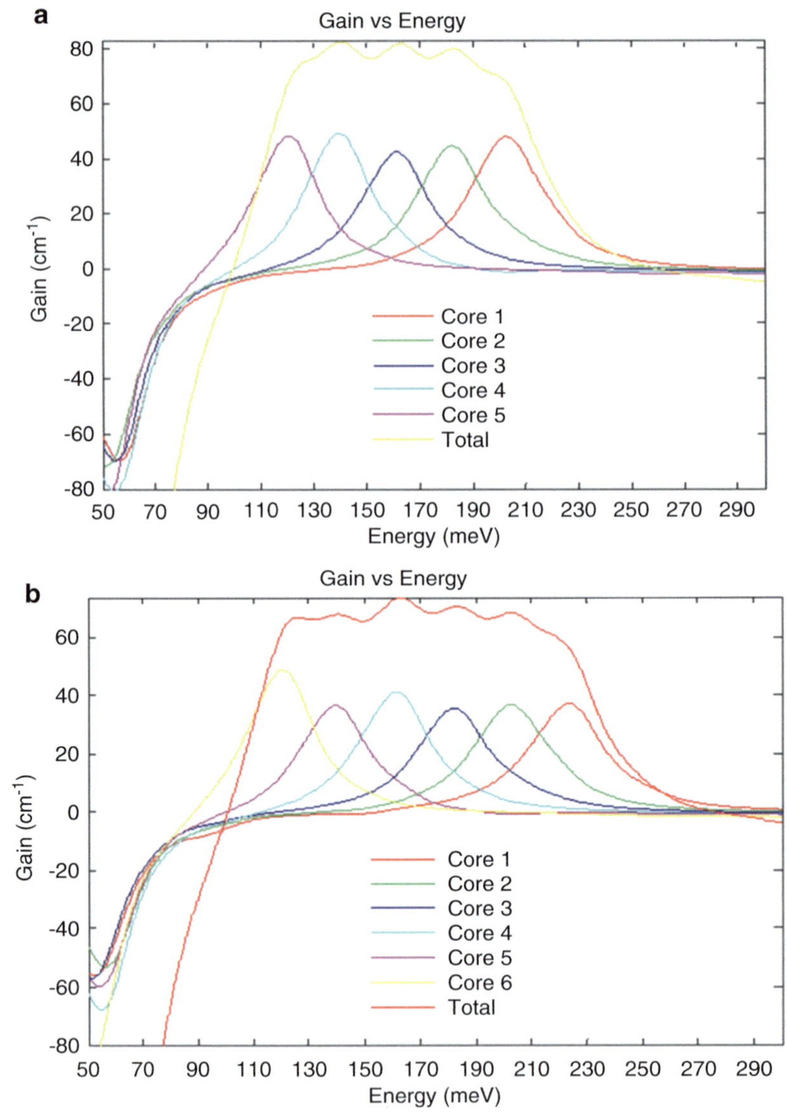

Fig. 16.4 Modal gain of (**a**) 5 and (**b**) 6 stage HQCL at resonant field

16.2.5 Growth, Fabrication and Testing

The broadband structure was grown by gas source molecular beam epitaxy on a 4 μm thick low-doped InP buffer (doped to n ~ 2×10^{16} cm^{-3}) on top of a low-doped InP substrate (doped to n ~ $1\text{–}2 \times 10^{17}$ cm^{-3}). QC stages (doped similarly to n ~ 4.7×10^{16} cm^{-3}), were grown, starting from the buffer region.

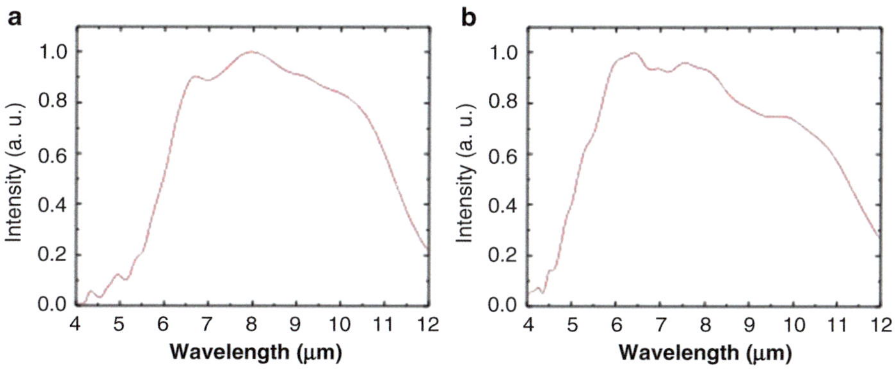

Fig. 16.5 Electroluminescence of (**a**) 5 and (**b**) 6 QC stage HQCL showing broad emission 5–12 μm

The growth of the active region was followed by a 100 nm thick InP (doped to $n \sim 2 \times 10^{16}$ cm^{-3}) and a 750 nm thick lattice matched $Ga_{0.47}In_{0.53}As$ grating layer.

A distributed feedback (DFB) grating array, with a total of 24 lasers were patterned on the grating layer by e-beam lithography. Etch depth of the gratings were 375 nm, with the grating period varying with the required Bragg wavelength. After standard grating fabrication [9], a 4 μm thick InP (doped to $n \sim 2 \times 10^{16}$ cm^{-3}) cladding layer and a 1 μm thick highly doped InP (doped to $n \sim 5 \times 10^{18}$ cm^{-3}) cap layer were grown on top of the grating by metal organic chemical vapor deposition. Standard double channel processing was followed to fabricate the DFB laser array, with the exception that Si_3N_4 was substituted for SiO_2 as an insulator [10]. Laser bars of cavity length 3 mm were cleaved and an AR coating of ~1.4 μm of Y_2O_3 was deposited on the both the facets. DFB array dies were then bonded epilayer up with indium solder on copper sub mounts. Testing was done on a thermoelectric cooler stage at 298 K.

Spectra were measured with a Fourier transform infrared spectrometer (FTIR) with a resolution of 0.125 cm^{-1} in rapid scan mode. Pulsed mode testing was done with a pulse width of 200 ns and a duty cycle of 1% (i.e., a repetition rate of 50 kHz). Single mode DFB emission from AR coated devices between 6.2–9.9 μm and 5.9–10.9 μm for 5 and 6 QC stage HQCL, respectively, was obtained as shown in Fig. 16.6 (a) and (b). The normalized spectrum of each DFB laser appears as a vertical line.

16.3 Conclusion

In conclusion, a technique for design and growth of highly efficient strain balanced $Al_{0.63}In_{0.37}As/Ga_{0.35}In_{0.65}As/Ga_{0.47}In_{0.53}As$ emitting between 5 and 11 μm was demonstrated. These individual cores emitting at discrete wavelengths were incorporated within a single active region of a HQCL. Design consideration like number of periods of QC stages need to be taken into account to have a flat gain over a broad

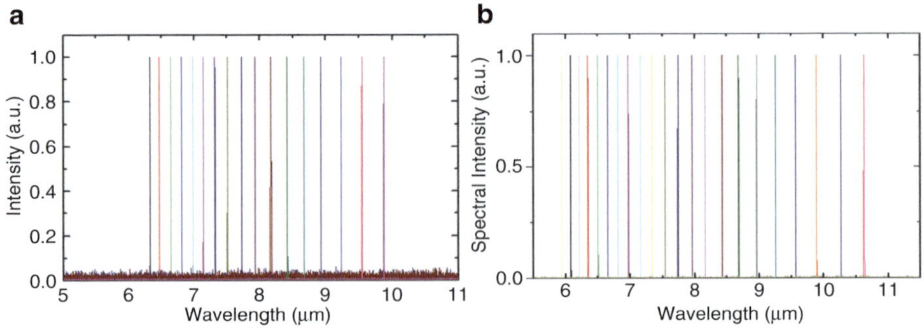

Fig. 16.6 Normalized emission spectrum for the AR coated HQCL DFBs at various wavelengths for (**a**) 5 and (**b**) 6 QC stage HQCL

wavelength range. Utilizing the DFB grating feedback HQCLs, having 5 and 6 QC stages are demonstrated to have emission between 6.2–9.9 μm and 5.9–10.9 μm respectively.

Acknowledgments This work is partially supported by the Department of Homeland Security Science and Technology Directorate (grant HSHQDC-13-C-00034), National Science Foundation (grants ECCS-1231289 and ECCS-1306397), Naval Air Systems Command (grant N68936-13-C-0124), and an Early Stage Innovations grant from NASA's Space Technology Research Grants Program. The authors would also like to acknowledge the encouragement and support of all the involved program managers.

References

1. Razeghi, M. High-Performance InP-Based Mid-IR Quantum Cascade Lasers. *Selected Topics in Quantum Electronics, IEEE Journal of* **15**, 941–951, doi:10.1109/JSTQE.2008.2006764 (2009).
2. Li, J. S., Chen, W. & Fischer, H. Quantum Cascade Laser Spectrometry Techniques: A New Trend in Atmospheric Chemistry. *Applied Spectroscopy Reviews* **48**, 523–559, doi:10.1080/05704928.2012.757232 (2013).
3. Furstenberg, R. *et al.* Stand-off detection of trace explosives via resonant infrared photothermal imaging. *Applied Physics Letters* **93**, 224103, doi:http://dx.doi.org/10.1063/1.3027461 (2008).
4. Bandyopadhyay, N. *et al.* Room temperature continuous wave operation of lambda λ ~ 3–3.2 μm quantum cascade lasers. *Applied Physics Letters* **101**, 241110, doi:Artn 241110 10.1063/1.4769038 (2012).
5. Bai, Y., Bandyopadhyay, N., Tsao, S., Slivken, S. & Razeghi, M. Room temperature quantum cascade lasers with 27% wall plug efficiency. *Applied Physics Letters* **98**, 181102, doi:Artn 18110210.1063/1.3586773 (2011).
6. Bandyopadhyay, N., Bai, Y., Slivken, S. & Razeghi, M. High power operation of λ ~ 5.2–11 μm strain balanced quantum cascade lasers based on the same material composition. *Applied Physics Letters* **105**, 071106, doi:http://dx.doi.org/10.1063/1.4893746 (2014).

7. Bandyopadhyay, N., Chen, M., Sengupta, S., Slivken, S. & Razeghi, M. Ultra-broadband quantum cascade laser, tunable over 760 cm-1, with balanced gain. *Optics Express* **23**, 21159–21164, doi:10.1364/OE.23.021159 (2015).
8. Hugi, A. et al. External cavity quantum cascade laser tunable from 7.6 to 11.4μm. *Applied Physics Letters* **95**, 061103, doi:http://dx.doi.org/10.1063/1.3193539 (2009).
9. Lu, Q. Y., Bai, Y., Bandyopadhyay, N., Slivken, S. & Razeghi, M. Room-temperature continuous wave operation of distributed feedback quantum cascade lasers with watt-level power output. *Applied Physics Letters* **97**, 231119–231113 (2010).
10. Bai, Y., Slivken, S., Darvish, S. R. & Razeghi, M. Room temperature continuous wave operation of quantum cascade lasers with 12.5% wall plug efficiency. *Applied Physics Letters* **93**, 021103, doi:http://dx.doi.org/10.1063/1.2957673 (2008).

Chapter 17
AlGaAs/GaAs Terahertz Quantum Cascade Laser with Gold-Based Metal – Metal Waveguide

Anna Szerling, Kamil Kosiel, Piotr Prokaryn, Michał Szymański, Artur Trajnerowicz, Maciej Sakowicz, Piotr Karbownik, Mariusz Płuska, Michał Walczakowski, and Norbert Pałka

Abstract In this chapter, we report the fabrication of $Al_{0.15}Ga_{0.85}As/GaAs$ ~3 THz frequency quantum cascade laser (QCL). Its operation is based on 3-quantum-well (3QW) modules, where the GaAs QWs are separated by $Al_{0.15}Ga_{0.85}As$ barriers. The laser's active region is build by stacking the 228 modules. The lasers were fabricated as gold-gold waveguide devices and were gold-gold mounted on GaAs receptor wafers. The 300 nm Au/5 nm Ti claddings were made by sputtering, for which we used 8×10^{-3} mbar pressure level and 50 W generator power. The wafer-bonding process was performed in the temperature range 350–400 °C. The structures were wet etched into ridges with widths in the range 50–160 microns and lengths 1.8 mm. The lasers operated with threshold current densities ~650 A/cm^2 at 77 K, when fed by 100–300 ns current pulses supplied with 300–1000 Hz repetition frequencies.

Keywords Terahertz quantum cascade laser • AlGaAs/GaAs structures • Metal – metal waveguides • Gold-gold mounting • Processing

THz radiation has a very broad spectrum of already existing and potential applications, ranging from measurements and diagnostics applied in fields such as molecular spectroscopy and three-dimensional imaging for biology, medicine, military

A. Szerling (✉) • K. Kosiel • P. Prokaryn • A. Trajnerowicz • M. Sakowicz • P. Karbownik M. Płuska
Institute of Electron Technology, Al. Lotników 32/46, 02-668 Warsaw, Poland
e-mail: szerling@ite.waw.pl

M. Szymański
Faculty of Applied Informatics and Mathematics, Warsaw University of Life Sciences, Nowoursynowska 159, 02-776 Warsaw, Poland

M. Walczakowski • N. Pałka
Military University of Technology, Kaliskiego 2, 00-908 Warsaw, Poland

M.F. Pereira, O. Shulika (eds.), *THz for CBRN and Explosives Detection and Diagnosis*, NATO Science for Peace and Security Series B: Physics and Biophysics,
DOI 10.1007/978-94-024-1093-8_17

and public safety purposes to telecommunications in free-space data transmission. Terahertz quantum cascade lasers (THz QCLs) can be a very good source for these applications. But still, even though around 15 years have passed since the creation of a first THz QCL, the highest operating temperatures for THz QCLs with operations not based on difference frequency generation effect do not exceed ~200 K [1]. There are two most popular waveguide types for THz QCLs: the semi-insulating surface plasmon waveguide, which suffers from relatively high free-carrier losses occurring in the highly doped semiconductor layer, and the metal-metal waveguide [2–4]. The latter one is equipped with two metallic claddings, one on each side of the active region. As the THz skin depth effect only reaches a few tens of nanometers into metal, the metal-metal outlay allows the fabrication of relatively low-loss waveguides with confinement factor approaching unity. Even in metal-metal waveguide THz QCLs, though, 30–70% of the waveguide loss can be associated with the claddings, with a specific value depending on emission wavelength, active region design and its doping [5]. Further improvement to the maximum THz QCL operating temperature through the optimization of both the active region and the claddings is likely. Optical losses linked with the metal claddings depend on type of the metal used [5], with just a few types of metal claddings tried out and applied to date with regard to THz quantum cascade lasers.

In this chapter, focusing on processing technology, we describe the $Al_{0.15}Ga_{0.85}As/GaAs$ quantum cascade laser (QCL) equipped with metal-metal waveguide based on gold, operating with ~3THz frequency. Its operation is based on 3-quantum-well (3QW) modules with the layer thickness sequence **4.3**, 8.9, **2.46**, 8.15, **4.1**, 16, given in nm, where the GaAs QWs are separated by $Al_{0.15}Ga_{0.85}As$ barriers bolded in the text. Only 5 nm-thick regions placed in the middle of the 16 nm-thick phonon wells were homogenously n-type doped with silicon, with planar concentration $3 \times 10^{10} cm^{-2}$. The laser's active region is build by stacking the 228 modules. The heterostructure was grown by molecular beam epitaxy technique in Riber Compact 21 machine on nominally (100) semiinsulating AXT-supplied epi-ready GaAs substrates at the temperature 580 °C. The lasers were fabricated as gold-gold waveguide devices and were gold-gold mounted on GaAs receptor wafers according to the scheme presented in Fig. 17.1.

Au-based waveguides have already proved to ensure relatively low waveguide losses [2–4], as well as high THz QCL T_{max} and low threshold current densities [6]. What is important, the fabrication of Au-based metallic claddings is much less demanding than, for instance, the Cu-based cladding technology, for which there is a problem of possible oxidation. For Ag-based claddings there are problems with wire bonding. Our studies covered the influence of technological parameters such as the thicknesses of metallic layers, process temperatures and durations on the properties of the resulting test structures [2]. As a first step that could offer valuable input towards the desired technology, we performed theoretical investigation of Au-based metal-metal waveguides designed for THz QCL. With the use of Drude-Lorenz model, the optical properties of constituent materials were checked. For the calculations we used a simplified model of the active region with a single uniform layer of total thickness equal to one of the active region mentioned above and the

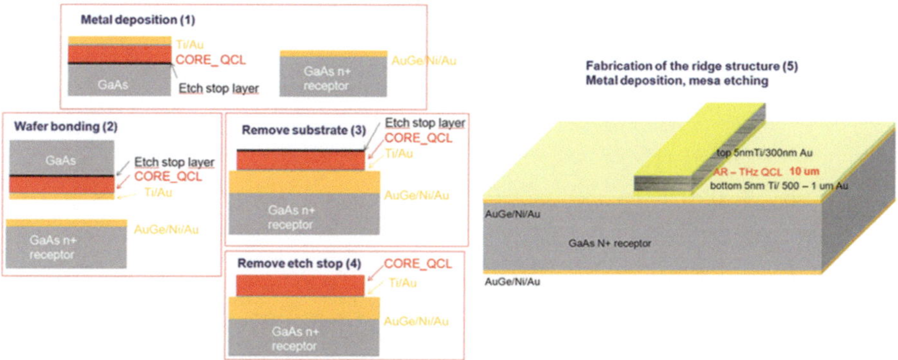

Fig. 17.1 The scheme of the processing of THz QCLs with metal – metal waveguides and the scheme of resulting laser chip

composition and doping concentration average for the host material. In this case, the THz QCL can be approximated as a 1-dimensional planar waveguide supporting the fundamental TM mode [7]. The complex refractive index values of constituent materials are calculated according to the Drude-Lorenz model [8, 9]. Solving the Helmholtz equation by means of the transfer matrix method allowed the calculation of waveguide losses for lasers with varied cladding materials (Table 17.1). A wider elaboration regarding metal-metal waveguides in THz QCLs that covers a wider array of metals is available in [2–4].

From our calculations it can be also concluded that the penetration depth of THz radiation in the Au layer should be below 100 nm. Thus we conclude that 300-nm thick Au layers fabricated in our waveguides are thick enough. It is also necessary to add the titanium layer between the semiconductor and gold layer in order to improve the gold adhesion and at the same time to block diffusion of gold into the active layer. Calculations show that up to about 30 nm of Ti does not significantly impact the waveguide losses in THz QCLs [2]. Taking into account theoretical predictions we deposited 5 nm Ti/300 nm Au as the cladding layers. The scheme of processing technology is presented in Fig. 17.1. The claddings were made by sputtering, for which we used 8×10^{-3} mbar pressure level and 50 W generator power. The technological parameters applied for the fabrication of the bottom Ti/Au cladding and for wafer bonding were selected on the basis of experimental work presented in [2]. Ti/Au-cladding fabrication parameters were determined to produce reliable metal-metal waveguide and at the same time low-resistivity stable metal-semiconductor contact. The fabricated contact was non-ohmic, however. Then the connection of THz QCL with receptor substrate was performed. The Au- Au bonding process was performed in the temperature range 350–400°C. The example of Au-Au bonded test structures is presented in Fig. 17.2.

After wafer bonding the substrate was mechanically removed by polishing and lapping down to about 50–70 μm. Then the rest of it was completely removed by selective wet etching. Finally, selective wet etching of the etch-stop AlAs epitaxial

Table 17.1 Theoretical results for the THz QCL ($\lambda = 100$ μm) with infinitely thick metal claddings made of titanium and gold layers as well as 300 nm-thick gold with a 5 nm-thick Ti spacer [2]

Metal	n_{metal}	α_{wg} (cm^{-1})
Ti (∞)	57.4 + i 65.9	78.10
Au (∞)	188.3 + i 237.4	48.69
5 nm-Ti/300 nm-Au		48.75

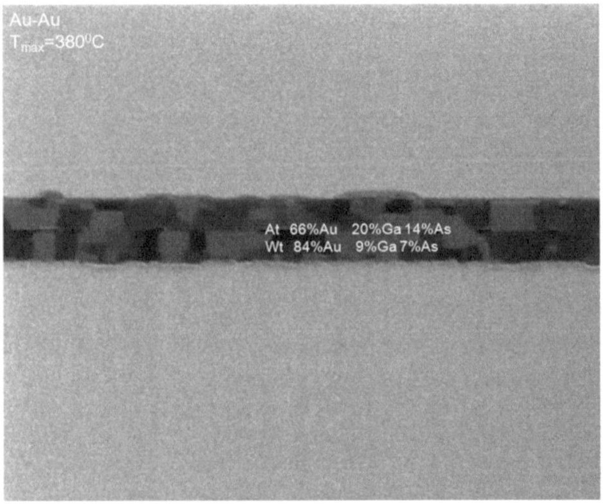

Fig. 17.2 SEM images of bonded GaAs/5 nmTi/300 nmAu and Au 300 nm/Ti 5 nm/GaAs test structures

layer was done. Substrate wet removal was performed in citric acid (50%) and hydrogen peroxide solution with 4:1 volumetric proportions, maintained at 22°C and with continuous stirring that ensured ~8 nm/s etching rate. The AlAs etch-stop layer was etched in HF:H_2O with 1:31 volumetric proportions and etching duration in the range of 10–20 s. In the next step the ridge structures were fabricated by photolithography and mesa wet etching, and finally top Ti/Au claddings were sputtered. The top claddings were fabricated using the same conditions as were used for the bottom ones with an additional lift-off process, however. The particular attention was paid to obtaining low roughness of mesa sidewalls [2]. Compositions of all solutions tested by us were taken from Ref [10]. The best solution found was H_3PO4:H_2O_2:H_2O (1:1:1), resulting in fast enough etching rate, and ensuring high etching uniformity in the plane of the wafer as well as smooth mesa sidewalls. The structures were wet etched into ridges with widths in the range 50–160 microns and lengths 1.8 mm. The fabricated devices were soldered ridge side up on a copper mount and wire bonded. Afterwards they were mounted on a cold stage in a vacuum cryostat and fed by 100–300 ns current pulses supplied with 300–1000 Hz repetition frequencies. The threshold current densities were around 650 A/cm^2 at 77 K (Fig. 17.3).

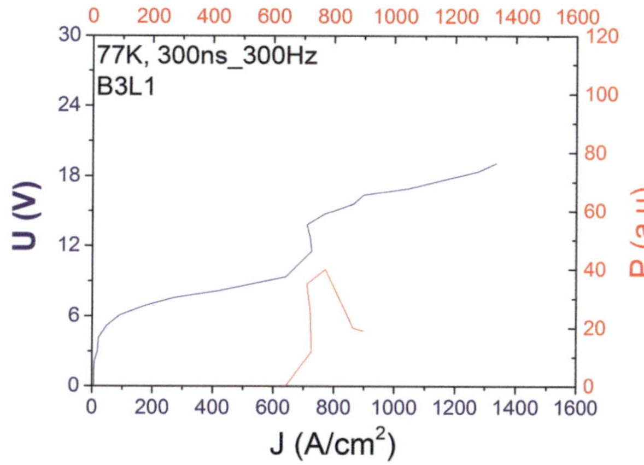

Fig. 17.3 The $Al_{0.15}Ga_{0.85}As/GaAs$ THz QCL basic characteristics

Acknowledgments The work was supported by Polish National Science Center by project 5028/B/T02/2011/40, project 2011/03/D/ST7/03146, project 2012/07/D/ST7/02568 and by COST ACTION MP1204 TERA-MIR Radiation: Materials, Generation, Detection and Applications.

References

1. S. Fathololoumi, E. Dupont, C. W. I. Chan, Z. R. Wasilewski, S. R. Laframboise, D. Ban, H. C. Liu, Terahertz quantum cascade lasers operating up to$\sim$ 200 K with optimized oscillator strength and improved injection tunneling, Optics Express, 20(4), 3866–3876, (2012). doi:10.1364/OE.20.003866
2. A. Szerling, K. Kosiel, M. Szymański, Z. Wasilewski, K. Gołaszewska, A. Łaszcz, M. Płuska, A. Trajnerowicz, M. Sakowicz, M. Walczakowski, N. Pałka, R. Jakieła, A. Piotrowska, Processing of AlGaAs/GaAs quantum-cascade structures for terahertz laser, Journal of Nanophotonics, 9 (1), 093079 (2015). doi: 10.1117/1.JNP.9.093079
3. M. Szymański, A. Szerling, K. Kosiel, Theoretical investigation of metal-metal waveguides for terahertz quantum-cascade lasers, Optical and Quantum Electronics, 47 (4), 843, (2014). doi: 10.1007/s11082-014-0007-z
4. M. Szymański, A. Szerling, K. Kosiel, M. Pluska, A study of different metals employed in metal-metal waveguides for terahertz quantum cascade lasers, Journal of Physics D: Applied Physics, will be published
5. M. Belkin, J. Fan, S. Hormoz, F. Capasso, S. Khanna, M. Lachab, A. Davies, E. Linfield, Terahertz quantum cascade lasers with copper metal-metal waveguides operating up to 178 K, Opt. Express 16, 3242–3248, (2008). doi:10.1364/OE.16.003242
6. S. Fathololoumi, E. Dupont, S. G. Razavipour, S. R. Laframboise, G. Parent, Z. Wasilewski, H. C. Liu, D. Ban, On metal contacts of terahertz quantum cascade lasers with a metal–metal waveguide, Semicond. Sci. Technol., 26, 1–5, (2011). doi:10.1088/0268-1242/26/10/105021
7. C. Sirtori, GaAs quantum cascade lasers: fundamentals and performance, EDP Sciences 7, 1–20, (2002). doi: 10.1051/bib-sfo:2002057
8. B. Ung, Y. Sheng, Interference of surface waves in a metallic nanoslit, Optics Express, 15(3), 1182–1190, (2007). doi: 10.1364/OE.15.001182
9. L. Quan-Yong, et al., Design of Low-Loss Surface-Plasmon Quantum Cascade Lasers, Japanese Journal of Applied Physics 48 12R, 122101, (2009). doi: 10.1143/JJAP.48.122101
10. A.R. Clawson, Guide to references on III$\pm$V semiconductor chemical etching, Materials Science and Engineering, 31, 1–438, (2001). doi:10.1016/S0927-796X(00)00027-9

Chapter 18
Infrared Laser Frequency Combs for Multispecies Gas Detection

M. Vainio, V. Ulvila, and L. Halonen

Abstract We discuss two methods for mid-infrared frequency comb generation in quadratic nonlinear materials. The first one of these methods is based on half-harmonic generation in a degenerate femtosecond optical parametric oscillator, which is synchronously pumped with a near-infrared laser frequency comb. The second method uses a continuous-wave-pumped optical parametric oscillator. In this approach, the quadratic nonlinearity is used not only for frequency down conversion, but also for the actual frequency comb generation from a continuous-wave pump laser field. Both methods can be used to produce coherent broadband mid-infrared light that is ideal for real-time spectroscopic detection of small traces of biomarkers, toxics, and other interesting molecular compounds. In particular, these methods provide access to the 3–5 μm molecular fingerprint region, which contains strong fundamental vibrational stretching bands of hydrocarbons.

Keywords Optical frequency comb • Laser frequency comb • Nonlinear optics • Optical parametric oscillator • Infrared spectroscopy

18.1 Introduction

An optical frequency comb (OFC) generator produces coherent broadband laser light that has a spectrum consisting of narrow equidistant lines. For an OFC suitable for high-resolution molecular spectroscopy, the comb tooth spacing is typically of the order of 100 MHz, and the number of comb teeth can be up to several hundred thousand. In the visible and near-infrared regions octave spanning frequency combs

M. Vainio (✉)
Molecular Science, Department of Chemistry, University of Helsinki, Helsinki, Finland

Photonics Laboratory, Tampere University of Technology, Tampere, Finland
e-mail: markku.vainio@helsinki.fi

V. Ulvila • L. Halonen
Molecular Science, Department of Chemistry, University of Helsinki, Helsinki, Finland

M.F. Pereira, O. Shulika (eds.), *THz for CBRN and Explosives Detection and Diagnosis*, NATO Science for Peace and Security Series B: Physics and Biophysics,
DOI 10.1007/978-94-024-1093-8_18

are routinely produced by mode-locked femtosecond lasers followed by additional spectral broadening in a highly nonlinear optical fiber. These combs can be fully stabilized by locking the comb offset frequency and repetition rate (tooth spacing) to a stable microwave frequency reference. The invention of fully stabilized OFCs in 1999 has led to several new high-accuracy laser spectroscopy methods [1]. As an example, dual-comb spectroscopy provides outstanding speed, accuracy, and spectral resolution in simultaneous detection of multiple molecular species [2, 3]. The discrete broadband nature of OFC light also allows for spectroscopy of molecules that have wide and complex spectra, and are therefore difficult to reliably detect with conventional lasers. Optical frequency combs can be combined with state-of-the-art laser spectroscopy techniques, such as cavity-enhanced spectroscopy, which allows parts-per-billion (ppb) level detection limits in real-time molecular spectroscopy [4, 5].

The mid-infrared region is in many cases the optimal spectral region for molecular spectroscopy, but OFCs in this region are scarce. Direct generation of OFCs at wavelengths longer than 2.8 μm by mode-locked lasers is yet to be demonstrated. Quantum cascade laser frequency combs have been investigated in the long mid-infrared (approximately 7–10 μm) and terahertz regions, while the 3–5 μm spectral window can be accessed by methods based on nonlinear optics [6, 7]. The basic idea of these methods is to use nonlinear frequency conversion to transfer a near-infrared OFC to the mid-infrared region. One of the best approaches for this is synchronously pumped degenerate femtosecond optical parametric oscillator (OPO), which is also referred to as half-harmonic generator [8–10]. The degenerate femtosecond OPO produces a mid-infrared frequency comb that is inherently phase-locked to the pump comb. The instantaneous spectral bandwidth of a degenerate fs-OPO can be hundreds of nanometers, and the average output power tens of milliwatts. A more detailed description and the results of recent research of our group on this method are given in Sect. 18.3.

In Sect. 18.4, we focus on so-called CQN (Cascaded Quadratic Nonlinearity) comb generation, which is a new nonlinear optics approach for frequency comb generation [11–13]. While the other quadratic nonlinear optics methods work merely as frequency down-converters that require a femtosecond OFC as an input, the CQN method uses cascaded quadratic nonlinearities to actually *generate* the frequency comb from a continuous-wave pump field. This new solution is suitable for direct mid-infrared frequency comb generation, but it can also be combined with nonlinear down-conversion, in which case a simple near-infrared CW pump laser can be used to produce a comb in the 3–5 μm region [14]. Here, we demonstrate generation of a composite CQN comb with a single instrument from blue to mid-infrared. The mid-infrared output power is up to several watts, which exceeds the power levels of most other infrared OFCs by one to three orders of magnitude.

18.2 Optical Parametric Oscillator (OPO)

An optical parametric oscillator (OPO) is a frequency down-converter, which generates two output frequencies, signal (ν_s) and idler (ν_i), in a nonlinear crystal when pumped with an intense laser beam at frequency ν_p. The crystal is placed inside an optical resonator, which is configured either as singly resonant (ν_s or ν_i resonating), or doubly resonant (both ν_s and ν_i resonating). The OPO process needs to conserve both photon energy and momentum, which respectively leads to two equations that both need to be fulfilled:

$$\nu_i = \nu_p - \nu_s, \tag{18.1}$$

$$\Delta k = k_p - k_s - k_i = 0, \tag{18.2}$$

where $k_x = 2\pi n_x/\lambda_x$ is the angular wavenumber, with n_x being the refractive index of the crystal material at the respective wavelength λ_x.

Equation 18.2 is often referred to as phase matching condition. Phase matching for generation of mid-infrared light at wavelength $\lambda_i = c/\nu_i \sim 3$ μm with near-infrared pumping is typically obtained by quasi-phase matching, for example by using a periodically poled lithium niobate crystal (PPLN).

18.3 Synchronously-Pumped Degenerate Femtosecond OPO

When an OPO is pumped with a mode-locked femtosecond laser whose output spectrum is a frequency comb, also the generated signal and idler spectra are combs [8]. A special case is a degenerate fs-OPO, where the signal and idler spectra overlap, such that they become indistinguishable via mutual injection locking (Fig. 18.1). The degenerate fs-OPO thus works as a half-harmonic generator. It has been shown that the output of a degenerate fs-OPO is inherently phase-locked to the pump comb, with two alternative, discrete states [8].

Figure 18.1 shows schematically the experimental setup of the degenerate fs-OPO developed in our laboratory. The OPO is pumped with 450 mW average power from a fully-stabilized Er-fiber laser comb (Menlo Systems GmbH), which has a − 3 dB bandwidth of about 50 nm centered at 1560 nm. The repetition frequency of pump pulses is 250 MHz, and the pulse length < 300 fs. The pump pulses are tightly focused in a 1-mm long Brewster-cut MgO-doped PPLN crystal, which is placed inside the OPO cavity. The cavity mirrors are gold-coated, except for the input coupler, which is a dielectric mirror that transmits the pump beam but is highly reflective for the signal/idler mid-infrared wavelengths. A thin pellicle beam splitter is used as a partial (~30%) output coupler for the mid-infrared beam.

Owing to the short duration of the pump pulses, the free spectral range of the OPO cavity needs to be synchronized with the pump repetition frequency, such that

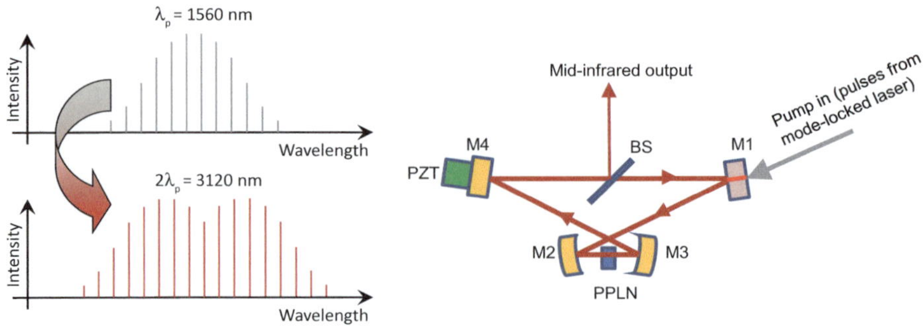

Fig. 18.1 *Left*: The principle of comb generation by degenerate fs-OPO that divides the center wavelength of the pump comb by two. In reality, the number of comb teeth is tens of thousands. *Right*: A schematic illustration of the fs-OPO cavity. Mirror M1 is the pump input coupler, and mirrors M2-M4 are gold-coated mirrors. Up to 70 mW of the resonating mid-infrared beam is coupled out with a beam splitter (BS). A piezo-electric actuator (PZT) is used for cavity length fine tuning and stabilization

the mid-infrared pulses resonating in the cavity overlap with consecutive pump pulses at each round trip. This condition sets the cavity length of our OPO to $c/250\ \text{MHz} = 1.199$ m. A piezoelectric actuator attached to one of the cavity mirrors is used to fine-tune the cavity length with a resolution better than 1 μm, which is needed to match the resonance condition of the synchronously pumped degenerate OPO. We use active locking to stabilize the OPO cavity length for long-term operation of several hours. The average mid-infrared output power of the stabilized fs-OPO is between 10 and 70 mW, depending on the pump power and output coupler used in the experiments.

The output spectrum of the degenerate fs-OPO can be modified by fine-tuning the OPO cavity length [9, 10]. The fully degenerate state, in which also the comb offset frequency is tightly locked to the pump comb, is the most favorable for applications like dual-comb spectroscopy, which requires high phase coherence and stability. Two fs-OPO output spectra corresponding to such degeneracy are exemplified in Fig. 18.2. Absorption features of compounds in the laboratory air are seen, not just due to the absorption between the OPO and the spectrum analyzer, but also owing to the absorption inside the OPO cavity. In fact, the OPO cavity increases the effective absorption path length by a factor of 5–10, leading to a respective enhancement in gas detection sensitivity [15].

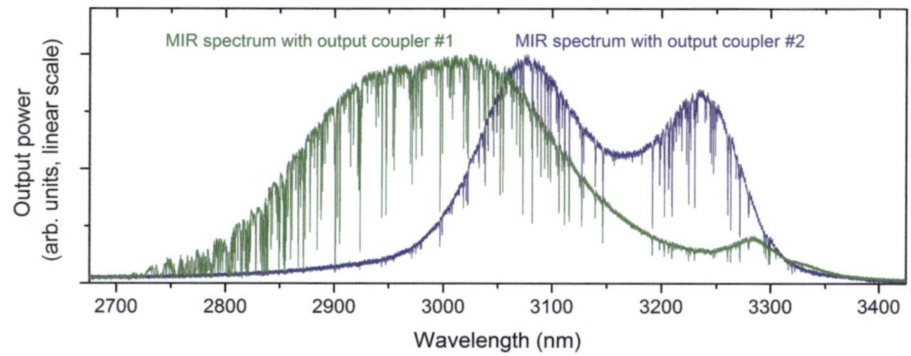

Fig. 18.2 Output spectrum of the fs-OPO measured with two different output couplers. Note that the comb tooth spacing (250 MHz, or ~8 pm) is too small to be resolved in the figure. The narrow peaks pointing downwards in the spectra are due to absorption of light in laboratory air over an effective absorption path length of 13 m. The absorption is mostly due to water vapor, but peaks resulting from ~2 ppm of CH_4 in air are also observed with a signal-to-noise ratio of >10 (limited by etalon effects). Both spectra were measured with a Fourier-transform infrared spectrometer (Bruker IFS 120 HR)

18.4 Continuous-Wave-Pumped CQN Comb Generation

The principle of optical frequency comb generation by cascaded quadratic nonlinearities is illustrated in Fig. 18.3 and explained in more detail in references [11–14]. In brief, the comb generation process from a CW pump field includes two steps: cavity-enhanced second-harmonic generation, and subsequent back conversion, where the back-converted light fills the pump resonator cavity modes while obeying the laws of energy and momentum conversion (Eqs. 18.1 and 18.2). So far the most accurate measurements of the CQN comb line spacing have confirmed the comb equidistance at 1 Hz level [13].

We have implemented the CQN scheme inside a CW-pumped OPO, which inherently transfers the frequency comb from near-infrared to mid-infrared [14]. In addition, residual nonlinear mixing processes extend the comb to higher frequencies, all the way to the blue, as demonstrated in Fig. 18.4. The composite frequency comb produced by the combination of various nonlinear processes covers three octaves, from 400 nm to 3.5 μm.

One of the advantages of the CQN comb is the possibility of achieving high, multiwatt output powers in the mid-infrared region. We have demonstrated up to 3 W of output power in the 3 μm fingerprint region [14], and more than 4 W in the 2 μm region [13] using CW near-infrared laser pumping. While the instantaneous bandwidth of the first CQN prototype mid-infrared comb is limited to approximately 10 nm, the center wavelength of the comb can easily be tuned over a wide range in the molecular fingerprint region, as shown in Fig. 18.5.

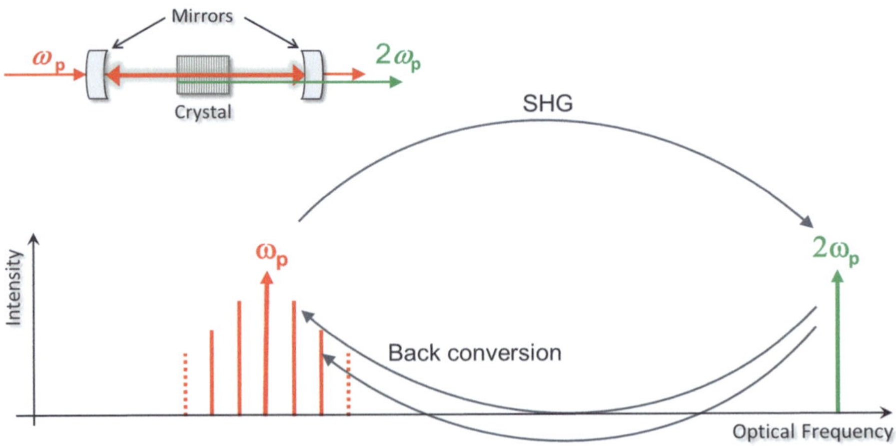

Fig. 18.3 The principle of CQN comb generation. A nonlinear crystal is placed inside an optical resonator. The crystal is designed for second harmonic generation (SHG) of the pump beam. With a properly adjusted phase-mismatch efficient back-conversion (cascading nonlinear effect) of the second harmonic light takes place, generating new frequency comb components close to the original pump frequency. The frequency separation of the comb teeth is determined by the length of the resonator

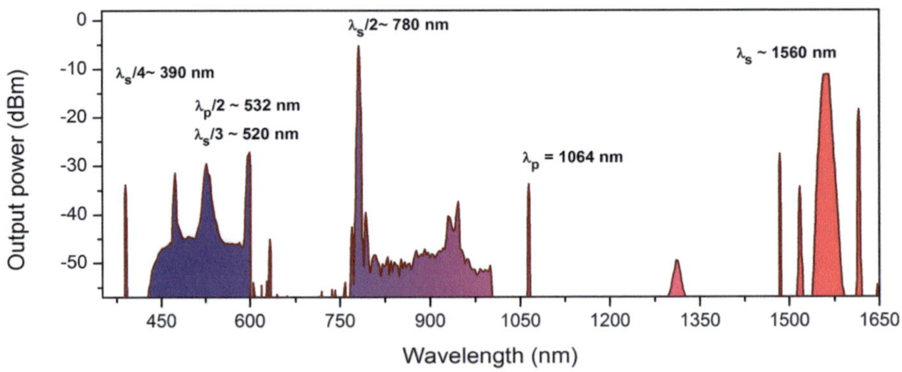

Fig. 18.4 Envelope spectrum of a composite CQN frequency comb generated by cascaded quadratic nonlinearities inside a CW OPO, which is pumped at 1064 nm [14]. Indexes p and s refer to OPO pump and signal wavelengths, respectively. The mid-infrared comb around the idler wavelength at >3 μm is outside the measurement range of the spectrum analyzer used here. Examples of the mid-infrared comb spectra, as measured with another spectrum analyzer, are shown in Fig. 18.5

18.5 Conclusion

In conclusion, we have reported two experimental approaches based on quadratic optical nonlinearities to generate optical frequency combs in the 3–5 μm molecular fingerprint region. The first approach relies on a degenerate, synchronously pumped

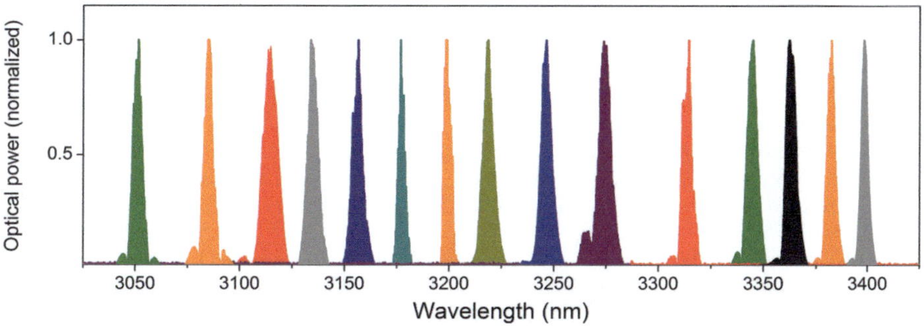

Fig. 18.5 Mid-infrared CQN comb envelope spectra measured with different phase-matching conditions of the CW OPO. Tuning of the comb center wavelength was done by changing the temperature and poling period of the OPO crystal [14]. The comb tooth spacing is not resolved in the figure

femtosecond OPO, which phase-coherently transfers a near-infrared frequency comb to the mid-infrared region. The instantaneous spectral coverage of the comb is hundreds of nanometers.

Our second approach for mid-infrared frequency comb generation is based on cascaded quadratic nonlinearities. This CQN method allows for comb generation practically at any wavelength from visible to mid-infrared with simple CW laser pumping. The output power can be up to several watts. In future, miniaturization of the CQN comb for field applications will be investigated [16].

Acknowledgments We thank the Academy of Finland, the Finnish Funding Agency for Innovation (Tekes), and the University of Helsinki for financial support.

References

1. S. A. Diddams, The evolving optical frequency comb. JOSA B 27, B51–B62 (2010). doi: 10.1364/JOSAB.27.000B51
2. S. Schiller, Spectrometry with frequency combs. Opt. Lett. 27, 766–768 (2002). doi: 10.1364/OL.27.000766
3. I. Coddington, W. Swann, N. Newbury, Coherent multiheterodyne spectroscopy using stabilized optical frequency combs. Phys. Rev. Lett. 100, 013902 (2008). doi: 10.1103/PhysRevLett.100.013902
4. F. Adler, M. J. Thorpe, K. C. Cossel, J. Ye, Cavity-Enhanced Direct Frequency Comb Spectroscopy: Technology and Applications. Annu. Rev. Anal. Chem. 3, 175–205 (2010). doi: 10.1146/annurev-anchem-060908-155248
5. A. Khodabakhsh, V. Ramaiah-Badarla, L. Rutkowski, A. C. Johansson, K. F. Lee, J. Jiang, C. Mohr, M. E. Fermann, A. Foltynowicz, Optical frequency comb spectroscopy at 3–5.4 μm with a doubly resonant optical parametric oscillator. arXiv:1603.09680 (2016).
6. A. Schliesser, N. Picqué, T. W. Hänsch, Mid-infrared frequency combs. Nat. Photonics 6, 440–449 (2012). doi:10.1038/nphoton.2012.142

7. M. Vainio, L. Halonen, Mid-infrared optical parametric oscillators and frequency combs for molecular spectroscopy. Phys. Chem. Chem. Phys. 18, 4266–4294(2016). doi: 10.1039/c5cp07052j
8. S. T. Wong, K. L. Vodopyanov, R. L. Byer, Self-phase-locked divide-by-2 optical parametric oscillator as a broadband frequency comb source. J. Opt. Soc. Am. B 27, 876–882 (2010). doi: 10.1364/JOSAB.27.000876
9. N. Leindecker, A. Marandi, R. L. Byer, K. L. Vodopyanov, Broadband degenerate OPO for mid-infrared frequency comb generation. Opt. Express 19, 6296–6302 (2011). doi: 10.1364/OE.19.006296
10. M. Vainio, M. Merimaa, L. Halonen, K. L. Vodopyanov, Degenerate 1 GHz repetition rate femtosecond optical parametric oscillator. Opt. Lett. 37, 4561–4563 (2012). doi: 10.1364/OL.37.004561
11. V. Ulvila, C. R. Phillips, L. Halonen, M. Vainio, Frequency comb generation by a continuous-wave-pumped optical parametric oscillator based on cascading quadratic nonlinearities. Opt. Lett. 38, 4281–4284 (2013). doi: 10.1364/OL.38.004281
12. I. Ricciardi, S. Mosca, M. Parisi, P. Maddaloni, L. Santamaria, P. De Natale, M. DeRosa, Frequency comb generation in quadratic nonlinear media. Phys. Rev. A 91, 063839 (2015). doi: 10.1103/PhysRevA.91.063839
13. V. Ulvila, C. R. Phillips, L. Halonen, M. Vainio, Spectral characterization of a frequency comb based on cascaded quadratic nonlinearities inside an optical parametric oscillator. Phys. Rev. A 92, 033816 (2015). doi: 10.1103/PhysRevA.92.033816
14. V. Ulvila, C. R. Phillips, L. Halonen, M. Vainio, High-power mid-infrared frequency comb from a continuous-wave-pumped bulk optical parametric oscillator. Opt. Express 22, 10535–10543 (2014). doi: 10.1364/OE.22.010535
15. M. W. Haakestad, T. P. Lamour, N. Leindecker, A. Marandi, K. L. Vodopyanov, Intracavity trace molecular detection with a broadband mid-IR frequency comb source. J. Opt. Soc. Am. B 30, 631–640 (2013). doi: 10.1364/JOSAB.30.000631
16. F. Leo, T. Hansson, I. Ricciardi, M. De Rosa, S. Coen, S. Wabnitz, M. Erkintalo, Walk-off-induced modulation instability, temporal pattern formation, and frequency comb generation in cavity-enhanced second-harmonic generation. arXiv:1510.04261 (2015).

Chapter 19
CBRN Defense Using THz Pulse Trains from Semiconductor Disk Lasers

Esa J. Saarinen, Jari Lyytikäinen, and Sanna Ranta

Abstract We propose THz generation from a photoconductive antenna illuminated by a train of optical pulses with a pulse repetition rate that corresponds to the desired THz frequency. This new method of THz generation can be seen as a hybrid between the conventional optical THz generation methods, where the optical source is either a heterodyne signal from two continuous wave lasers or a single short pulse. Therefore, the method holds promise for generating both coherent broadband and narrow-linewidth continuous wave THz radiation. The high-repetition rate optical pulse train is obtained from a semiconductor disk laser harmonically mode-locked by a semiconductor saturable absorber mirror and an intracavity etalon. Optical pulse trains with pulse repetition rates from 190 GHz to 580 GHz are demonstrated at an average optical output power of 1 W. This power level is enough for driving full arrays of photoconductive antennas. The approach may provide a compact and powerful THz source for CBRN defense.

Keywords Semiconductor disk laser • Vertical-external-cavity surface-emitting laser • Passive harmonic mode locking • Photoconductive antenna • THz generation • CBRN defense

19.1 Introduction

THz spectroscopy is a powerful tool for detecting hazardous substances and explosives [1]. Molecular motion at THz frequencies typically involves every atom in the molecule. As a result, the response of the molecule to THz radiation provides a unique signature by which the molecule can be identified [2].

THz sources can be categorized by the approach used for generating THz radiation. Typical approaches include THz generation directly from a laser [3, 4], up-conversion of the frequency from microwaves [5, 6], and down-conversion from optical frequencies [7]. Down-conversion from optical frequencies can be

E.J. Saarinen (✉) • J. Lyytikäinen • S. Ranta
Optoelectronics Research Centre, Tampere University of Technology, Tampere, Finland
e-mail: esa.saarinen@nanofoot.fi

M.F. Pereira, O. Shulika (eds.), *THz for CBRN and Explosives Detection and Diagnosis*, NATO Science for Peace and Security Series B: Physics and Biophysics,
DOI 10.1007/978-94-024-1093-8_19

accomplished by optical rectification [8] or photoconductive switching [7]. Common laser types used to optically generate THz radiation include pulsed solid state lasers, which are bulky, expensive and complicated [9], and continuous wave (CW) diode lasers [10], which work properly in a photomixer setup only when beams from two lasers are precisely combined and stabilized in frequency and polarization.

Semiconductor disk laser (SDL) is a laser type that can generate a low-divergence beam with high optical power at any wavelength enabled by the mature semiconductor growth technology. To date, the selection of wavelengths available from SDLs spans from 250 to 5 μm [11], the average output power of SDLs has been scaled to over 100 W [12], and single-frequency operation has been demonstrated with sub-20 kHz linewidths [13]. SDLs are also capable of generating ultrashort pulses via passive mode-locking [14]. In particular, the large gain cross section and short carrier lifetime in the semiconductor gain medium allow very high pulse repetition rates on the order of a hundred GHz to be achieved [15–17].

In this paper we propose a new method to produce THz radiation with a photoconductive antenna. This method can be seen as a hybrid of the conventional optical methods of THz generation: instead of either a single sub-picosecond pulse or a heterodyne signal from two continuous wave (CW) lasers, an optical pulse train with a pulse-to-pulse period on the order of one picosecond would drive the photoconductive antenna. The repetition rate of the pulses corresponds to the desired THz frequency. The new scheme may allow generating both a narrow-linewidth CW THz wave and broadband THz radiation, depending on the duty cycle and frequency content of the optical pulse train. The pulse train is generated with a passively mode-locked SDL capable of watt level output powers. The approach may help realize a room-temperature milliwatt THz source of a size of a shoebox.

The paper is organized as follows. In Sect. 19.2 we discuss the basic principles and potential advantages of generating THz radiation with SDLs operating at THz pulse repetition rates. Section 19.3 reviews our research for upscaling the repetition rate of passively mode-locked SDLs from 200 GHz to 600 GHz. The concluding remarks are given in the end.

19.2 THz Generation with a THz-Repetition Rate Optical Pulse Train

The basic principle of THz generation with a THz-repetition rate optical pulse train is illustrated in Fig. 19.1. An optical pulse train from an SDL with a pulse repetition rate corresponding to the desired THz frequency is directed at a photoconductive (PC) antenna. Photocurrent that drives the PC antenna follows the optical pump waveform from the SDL. Therefore, the photocurrent increases and decreases at the

Fig. 19.1 Illustration of THz generation with a THz-repetition rate optical pulse train

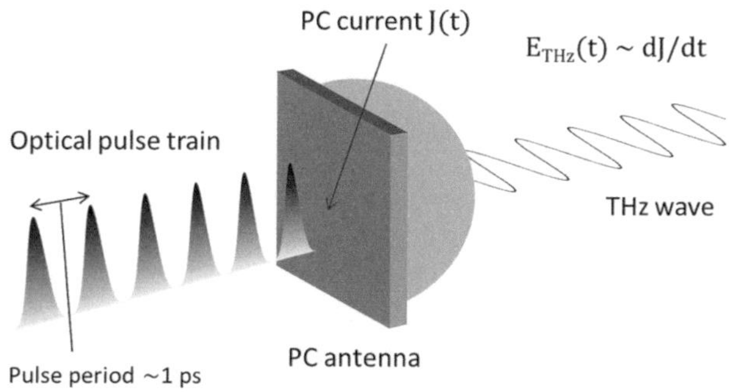

repetition frequency of the optical pulses. Changes in the photocurrent in time are proportional to the electromagnetic wave generated by the antenna.

The optical pulse trains generated by SDLs can be rather flexibly controlled. Namely, the pulse repetition rate and the pulse parameters such as pulse duration and chirp can be varied by tuning the cavity length, temperature of the cavity elements, and laser wavelength [18]. This allows the duty cycle and frequency content of the pulse train to be adjusted, which in turn gives access to controllable THz signal from the PC antenna. In principle, both coherent broadband and narrow-linewidth CW THz radiation could be generated. Furthermore, PC antennae optimized for optical pulse trains with different frequency contents could be used to take full advantage of the pump waveform flexibility provided by SDLs.

The method may permit very stable CW THz radiation to be produced. Compared to conventional photomixing with two stabilized lasers, the SDL pulse train enables the photocurrent to be induced by fully coherent, phase-locked optical modes, which spatially overlap perfectly, do not compete with each other and have identical polarization. Therefore, phase jumps of individual frequency components of the optical pump signal do not affect much the total collective waveform. The aforementioned properties of the pump beam may result in low noise of the THz signal with no need for complicated phase-lock loops.

In principle, SDL emission can be generated at wavelengths of 1 μm, 1.3 μm, 1.55 μm, or 2 μm, for example, to take advantage of different photoconductor materials. Several watts of average optical output power commonly available from SDLs allows driving full arrays of THz emitters and detectors without the need of additional amplifier chains. Furthermore, SDLs' excellent beam quality enables easy focusing and manipulation of the pump light for various PC antenna configurations.

19.3 Optical Pulse Trains from SDLs

Semiconductor disk lasers (SDLs) can be passively mode-locked by a semiconductor saturable absorber mirror (SESAM) [14]. Sub-picosecond pulses are routinely generated with proper gain element and SESAM design that takes into account dispersion and power dependent phase shifts in the semiconductor medium [19].

Large gain cross section and short gain relaxation time of semiconductors make SDLs almost free from Q-switching instabilities even at multi-GHz repetition rates [20]. Therefore, SDLs are intrinsically suitable for ultrashort pulse generation at high repetition rates and high average power. Indeed, repetition rates up to 200 GHz have been obtained from SDLs [15–17].

The highest pulse repetition rates are obtained via passive harmonic mode locking using a SESAM for the initiation of pulse operation and by coupling the laser cavity to an intracavity etalon. A high harmonic frequency (up to 450 times the fundamental cavity repetition rate) can be achieved when the laser cavity length is adjusted to match an integer multiple of the optical thickness of the intracavity etalon. The etalon can be a diamond heat spreader bonded onto the SESAM for improved heat dissipation.

Figure 19.2 shows a passively harmonically mode-locked SDL emitting at the wavelength of 1050 nm. The laser is similar to the SDL described in [17]. The gain structure was grown by molecular beam epitaxy and comprised a resonant periodic gain section based on InGaAs quantum wells (QWs) and a distributed Bragg reflector (DBR) with 26 layer pairs of GaAs/AlAs. The back side of the chip was bonded to a chemical vapor deposited diamond heat spreader. A 2-layer SiO_2/TiO_2 low dispersion antireflection coating was deposited on the chip to enable sub-picosecond pulse operation. The SESAM was also grown by molecular beam epitaxy and comprised a 28-layer pair GaAs/AlGaAs DBR and 2 InGaAs QWs in antiresonant subcavity. Ni ion bombardment was used to shorten the recovery time of absorption to a few ps. A diamond heat spreader was capillary bonded onto the SESAM for heat extraction and spectral filtering purposes.

Figure 19.3 shows how spectral filtering by the intracavity etalon results in harmonic mode locking. The longitudinal laser modes within the gain spectrum (plotted in red) encounter wavelength dependent loss with a periodicity determined by the free spectral range (FSR) of the etalon. The modes that coincide with transmission maxima prevail and can contribute to the formation of the mode-locked pulse train. The pulse repetition rate of the laser then equals the FSR of the etalon, instead of the mode spacing of the laser cavity. The length of the laser cavity has to match an integer multiple of the etalon optical thickness for perfect coupling of the two cavities. Moreover, high finesse of the etalon is preferred for efficient filtering of unwanted cavity modes. In practice, spherical bowing of the etalon due to imperfect polishing and the deviation from parallelism typically larger than 100 arc seconds are the main factors limiting the effective finesse of thin diamond etalons [21].

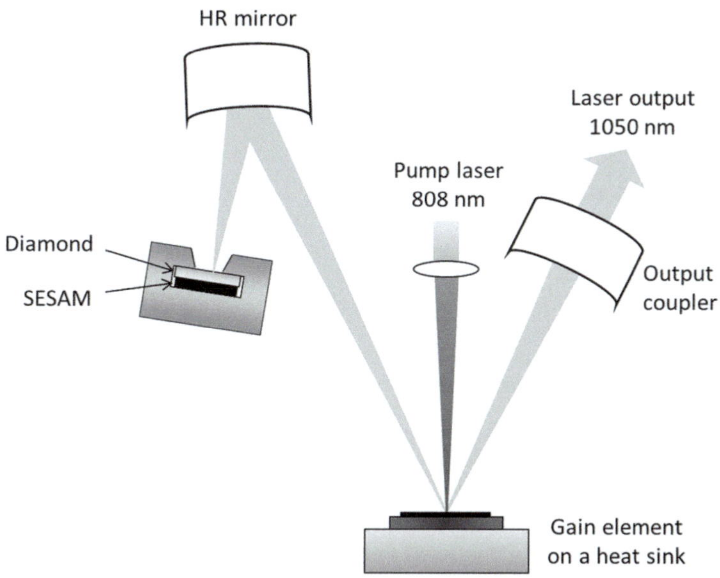

Fig. 19.2 Passively harmonically mode-locked SDL emitting at 1050 nm. *HR* high-reflective, *SESAM* semiconductor saturable absorber mirror

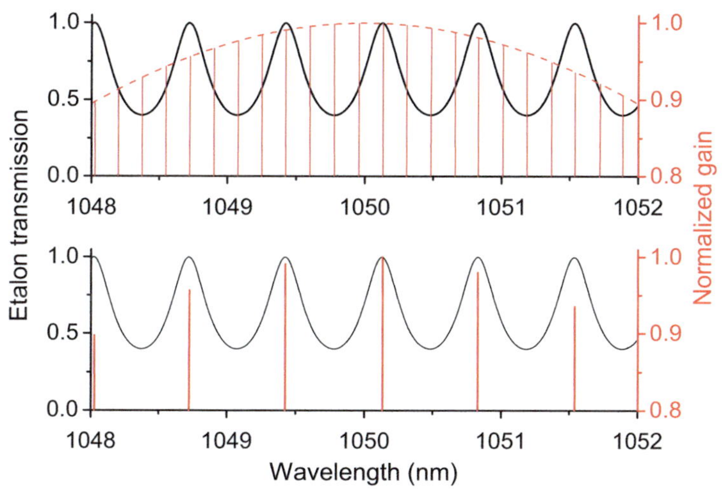

Fig. 19.3 Schematic representation of spectral filtering due to the diamond etalon placed inside the SDL cavity. The *black solid line* shows the etalon transmission. Gain spectrum is represented by the *red dashed line*. The *red vertical lines* illustrate longitudinal laser cavity modes. Modes that coincide with the transmission peaks of the etalon are favored, as shown in the lower panel

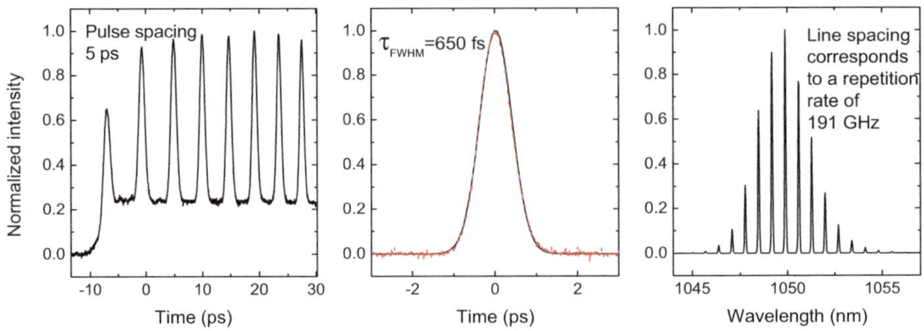

Fig. 19.4 Collinear second harmonic autocorrelation trace (*left*) and optical spectrum (*right*) for an SDL generating pulses at a repetition rate of 191 GHz. The *center panel* shows a close-up of the autocorrelation and a Gaussian fit to the data, indicating a pulse duration of 650 fs. The average optical output power of the laser was 0.8 W

Spectral filtering can be made easier by building a short laser cavity with wider frequency spacing between the cavity modes. This favors generation of a stable pulse train without peak power fluctuations at the laser cavity round trip frequency. The etalon should be in such a location in the cavity that transmission is preferred over reflection from the etalon for the light amplified in the laser resonator. Otherwise, frequencies in the wide stop bands of the etalon may be preferred over the frequencies in the narrow pass bands of the etalon, which prevents efficient wavelength selecting by the etalon.

Figures 19.4 and 19.5 show optical pulse trains with repetition rates of ~190 GHz obtained from SDLs similar to that depicted in Fig. 19.2. Characterization of the pulse trains includes collinear second harmonic autocorrelation and optical spectrum. The pulse durations given by Gaussian fittings to the data are 650 fs and 930 fs for the pulses shown Figs. 19.4 and 19.5, respectively. The demonstrated variation in pulse parameters and the corresponding differences in the duty cycles of the pulse trains were obtained mainly by changing the resonance characteristics of the gain element. The gain element for the pulse train in Fig. 19.5 had a stronger subcavity resonance formed by the bottom DBR and chip surface than that in Fig. 19.4.

If the etalon provoking harmonic mode locking is not placed at the laser cavity end, multiple cavities are formed and all of them have to be tuned with respect to the etalon optical thickness. Optimization of such an arrangement may be problematic and easily results in incomplete mode locking with a CW background in the autocorrelation of the pulse train. An example of this kind of behavior is shown in Fig. 19.6, where autocorrelation and spectrum are plotted for a 330 GHz pulse train obtained from an SDL harmonically mode-locked by a 250 μm thick YAG etalon placed inside the laser cavity with no contact to the end mirrors of the cavity.

Figure 19.7 shows autocorrelation measured for a 1 W optical pulse train with repetition rate of 585 GHz. This is the highest repetition rate obtained from SDLs to date. The pulse duration estimated from Gaussian fitting to the data was 440 fs. The

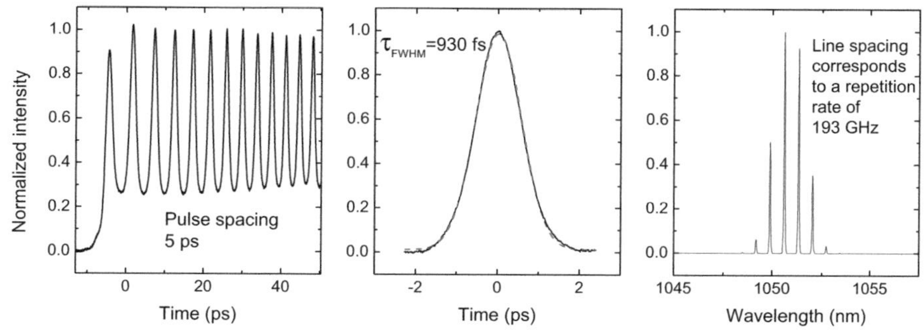

Fig. 19.5 Collinear second harmonic autocorrelation trace (*left*) and optical spectrum (*right*) for an SDL generating pulses at a repetition rate of 193 GHz. The *center panel* shows a close-up of the autocorrelation and a Gaussian fit, which gives a pulse duration of 930 fs. The average optical output power of the laser was 1.2 W

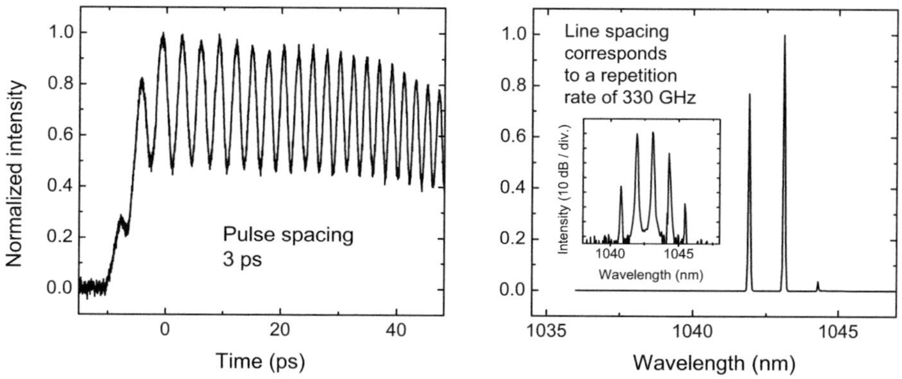

Fig. 19.6 Collinear second harmonic autocorrelation trace (*left*) and optical spectrum (*right*) measured for an SDL generating pulses at a repetition rate of 330 GHz. The inset in the *right panel* shows the spectrum in logarithmic scale

high repetition rate was obtained with a similar setup to that shown in Fig. 19.2. The repetition rate was three times the FSR of the diamond etalon bonded to the SESAM, indicating that lasing occurred only at every third transmission peak of the etalon. It should be noted, however, that such a harmonic mode locking state was hard to stabilize, rendering it impractical for real world applications. Future work will focus on finding means to better stabilize and control the pulse trains and scale up the SDL repetition rates beyond 1 THz.

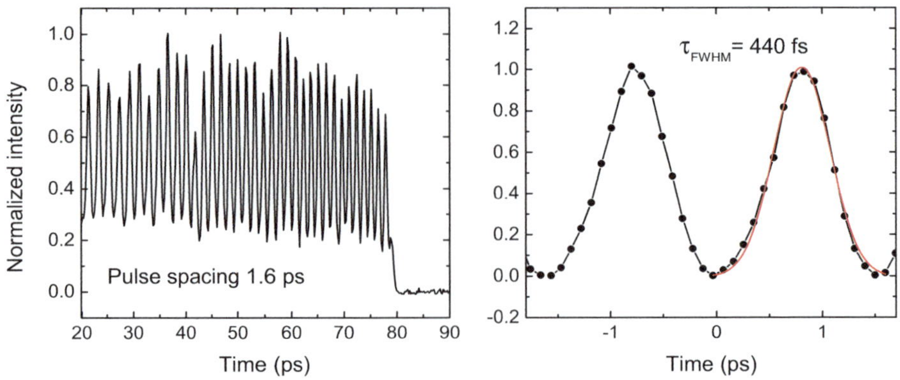

Fig. 19.7 Collinear second harmonic autocorrelation trace of an SDL generating pulses at a repetition rate of 585 GHz. The *right panel* shows a close-up of the autocorrelation with a Gaussian fit giving a pulse duration of 440 fs. The average optical output power of the laser was 1.0 W

19.4 Conclusion

THz-repetition rate optical pulse trains from SDLs were investigated. The potential using such pulse trains for generating both a highly stable CW THz waveform and coherent broadband THz radiation was discussed. The advantages of SDLs as optical pump sources for THz antennas were highlighted. Optical pulse trains were demonstrated from SDLs with repetition rates ranging from 0.2 THz to 0.6 THz. The watt level output power of the lasers allows driving very large area emitters or full arrays of PC antennas.

Acknowledgments This work was supported in part by the Academy of Finland under project T-REPTILE (269121).

References

1. J. F. Federici, B. Schulkin, F. Huang, D. Gary, R. Barat, F. Oliveira, and D. Zimdars, "THz imaging and sensing for security applications—explosives, weapons and drugs," Semicond. Sci. Technol. 20 S266–S280 (2005).
2. M. R. Leahy-Hoppa, M. J. Fitch, R. Osiander, "Terahertz spectroscopy techniques for explosives detection," Anal. Bioanal. Chem. 395, 247–257 (2009).
3. B. S. Williams, "Terahertz quantum cascade lasers," Nature Photon. 1, pp. 517–525 (2007).
4. Y.U. Jeong, G. M. Kazakevich, S. H. Park, K. Lee,B. C. Lee, Y.-H. Cha, P. Ahn, and J. Mun, "High power table-top THz free electron laser and its application," Infrared, Millimeter, and Terahertz Waves 2009, pp. 1–2 (2009).
5. N. Orihashi, S. Suzuki, and M. Asada, "One THz harmonic oscillation of resonant tunneling diodes," Appl. Phys. Lett., vol. 87, p. 233501 (2005).

6. J. Sharma and H. Krishnaswamy, "216- and 316-GHz 45-nm SOI CMOS Signal Sources Based on a Maximum-Gain Ring Oscillator Topology," IEEE Trans. Microw. Theory Tech. 61, 492–504 (2013).
7. S. Preu, G. H. Dohler, S. Malzer, L. J. Wang, and A. C. Gossard, "Tunable, continuous-wave Terahertz photomixer sources and applications," J. Appl. Phys. vol. 109, p. 061301 (2011).
8. Y.-S. Lee, T. Meade, V. Perlin, H. Winful, and T. B. Norris, "Generation of narrow-band terahertz radiation via optical rectification of femtosecond pulses in periodically poled lithium niobate," Appl. Phys. Lett. 76, 2505–2507 (2000).
9. F. Blanchard, G. Sharma, L. Razzari, X. Ropagnol, H.-C. Bandulet, F. Vidal, R. Morandotti, J.-C. Kieffer, T. Ozaki, H. Tiedje, H. Haugen, M. Reid,and F. Hegmann, "Generation of Intense Terahertz Radiation via Optical Methods," IEEE J. Sel. Top. Quantum Electron., vol. 17, pp. 5–16 (2011)
10. A. J. Deninger, T. Gobel, D. Schonherr, T. Kinder, A.l Roggenbuck, M. Koberle, F. Lison, T. Muller-Wirts, and P. Meissner, "Precisely tunable continuous-wave terahertz source with interferometric frequency control," Rev. Sci. Instrum. vol. 79, p. 044702 (2008).
11. O. G. Okhotnikov (Ed.), Semiconductor disk lasers: physics and technology, Wiley-VCH, Weinheim (2010).
12. B. Heinen, T.-L. Wang, M. Sparenberg, A. Weber, B. Kunert, J. Hader, S.W. Koch, J.V. Moloney, M. Koch and W. Stolz, "106 W continuous-wave output power from vertical-external-cavity surface-emitting laser," Electron. Lett. 48, 516–517 (2012).
13. A. Rantamäki, J. Rautiainen, A. Sirbu, A. Mereuta, E. Kapon, and O. G. Okhotnikov, "1.56 μm 1 watt single frequency semiconductor disk laser," Opt. Express 21, 2355–2360 (2013).
14. B. W Tilma, M. Mangold, C. A Zaugg, S. M Link, D. Waldburger, A. Klenner, A. S Mayer, E. Gini, M. Golling and U. Keller, "Recent advances in ultrafast semiconductor disk lasers," Light Sci. Appl. 4, e310 (2015).
15. A. H. Quarterman, A. Perevedentsev, K. G. Wilcox, V. Apostolopoulos, H. E. Beere, I. Farrer, D. A. Ritchie and A. C. Tropper, "Passively harmonically mode-locked vertical-external-cavity surface-emitting laser emitting 1.1 ps pulses at 147 GHz repetition rate," Appl. Phys. Lett. 97, 251101 (2010).
16. K. G. Wilcox, A. H. Quarterman, V. Apostolopoulos, H. E. Beere, I. Farrer, D. A. Ritchie, and A. C. Tropper, "175 GHz, 400-fs-pulse harmonically mode-locked surface emitting semiconductor laser.," Opt. Express 20, 7040–7045 (2012).
17. E. J. Saarinen, A. Rantamaki, A. Chamorovskiy and O. G. Okhotnikov, "200 GHz 1 W semiconductor disc laser emitting 800 fs pulses," Electron. Lett. 48, 1355–1357 (2012).
18. E. J. Saarinen, "Optical Pulse Generation with Semiconductor Disk Lasers," Ph.D. dissertation, Tampere University of Technology, Tampere (2012).
19. O. D. Sieber, M. Hoffmann, V. J. Wittwer, M. Mangold, M. Golling, B. W. Tilma, T. Südmeyer, and U. Keller, "Experimentally verified pulse formation model for high-power femtosecond VECSELs," Appl. Phys. B 113, 133–145 (2013).
20. U. Keller, A. C. Tropper, "Passively modelocked surface-emitting semiconductor lasers," Phys. Report 429, 67 –120 (2006).
21. A. R. Martel and A. Fullerton, "An Introduction to the TFI Etalon," Space Telescope Science Institute, Technical report JWST-STScI-002059, SM-12 (2010).

Chapter 20
Progress in Development of the Resonant Tunneling Diodes as Promising Compact Sources at the THz Gap Bottom

Andres Udal, Martin Jaanus, Gintaras Valušis, Irmantas Kašalynas, Zoran Ikonic, and Dragan Indjin

Abstract In recent years remarkable progress has been made in filling the "terahertz gap" of compact sources from the "optical" side using quantum cascade lasers and difference frequency generation systems. On the "electronic" side the range up to 0.5 THz is covered by several devices like Gunn diodes, IMPATT diodes and HEMT transistors. However for the most difficult central range 0.5–2 THz, which includes frequencies for detection of explosives and several airborne threats, along with the popular semi-compact Schottky diodes based frequency multiplier systems, the resonant tunneling diodes (RTD) as micrometer-size range ultrafast electronic devices able to operate at room temperature may offer a real solution for the design of compact portable equipment. It has been predicted that frequency limit for operation of the RTDs may extend up to 3 THz if the problem of the low, microwatt-range output power could be solved. Over the last 10 years remarkable progress has been achieved in increasing the output power of RTDs by almost 2 orders of magnitude, and in extending the operation frequencies from earlier 0.7 THz range to the values near 2 THz, thus making RTDs competitive with the Schottky diode based multipliers. The chapter compares the RTDs with other compact THz sources and discusses the design approaches that have yielded remarkable increase of power and frequency of RTDs.

Keywords Terahertz • Terahertz gap • Compact source • Resonant tunneling • Resonant tunneling diode • RTD • Slot antenna

---

A. Udal (✉) • M. Jaanus
Department of Computer Control, Tallinn University of Technology, Tallinn, Estonia
e-mail: andres.udal@ttu.ee

G. Valušis • I. Kašalynas
Department of Optoelectronics, State Research Center for Physical Sciences and Technology, Vilnius, Lithuania

Z. Ikonic • D. Indjin
School of Electronic and Electrical Engineering, University of Leeds, Leeds, UK

M.F. Pereira, O. Shulika (eds.), *THz for CBRN and Explosives Detection and Diagnosis*, NATO Science for Peace and Security Series B: Physics and Biophysics,
DOI 10.1007/978-94-024-1093-8_20

20.1 Introduction to the THz Compact Sources

Terahertz frequency range of EM radiation, usually associated with one decade above and below 1 THz (wavelength 0.3 mm), is the challenging boundary domain between microwave radio electronics and far infrared optics. As this range covers the vibrational and rotational frequencies of molecules, a large number of high priority spectroscopic and imaging applications exist in security, chemical remote sensing, astronomy, biomedical diagnostics etc. [1–7]. Since THz frequencies are too high for most of the conventional electronic devices based on drift-diffusion transport of charge carriers, and at the same time the EM radiation quanta $h\nu$ are too small in energy (compared to thermal energy kT) to realize easily the population inversion necessary for lasing, the center of THz range is often called "terahertz gap" (see below Fig. 20.1), expressing the lack of effective compact sources [1–7]. In the present study we focus on development of compact semiconductor devices for this THz gap range, leaving aside the bulky vacuum devices like free electron lasers, backward wave oscillators, gas lasers etc. [5, 6].

Considering the optical side of the THz domain in Fig. 20.1, the most important two compact source devices are the quantum cascade lasers (QCL), which need cryogenic cooling below 200 K, and difference frequency generation (DFG) devices that in modern compact realization use intracavity mixing of two mid-IR fields generated by the dual-wavelength mid-IR QCLs and can operate at room temperature (RT) but still with below 0.1% range plug-in efficiency [8, 9]. Although in recent years the pulsed power of THz QCLs has reached the impressive 1 W level (at liquid helium temperatures) [10–13], thus increasing ~1000 times the values given in earlier THz gap figures (Fig. 20.1), the problems still include the need for cooling below 200 K for CW operation and rather limited frequency range 1.2–5 THz [1–5, 10–12]. The lower frequency limit comes from the decreasing photon energy in comparison with the thermal energy and upper limit is defined by the optical phonon energy 36 meV, causing the opaque "restrahlen band" for the commonly used material system GaAs/AlGaAs. It has been estimated that for GaAs/AlGaAs the theoretical limit of CW operation is 240 K [4], if not considering the technologically complicated quantum dot solutions and/or strong magnetic fields [3–6]. The mentioned limitations of QCL and DFG based sources emphasize the need for further search of alternative approaches.

Considering the electronic side of the THz domain in Fig. 20.1, there are different compact semiconductor sources available for frequencies below 0.3–0.5 THz, for example, the Gunn diodes and the impact ionization avalanche transit time (IMPATT) diodes [2–6]. Also, the Si-MOSFET transistors and GaAs pHEMTs (pseudomorphic high electron mobility transistors) are capable to reach frequencies 0.3–0.4 THz in the case of nanoscale gate lengths, below 100 nm [14]. Furthermore, the InP-HEMTs (indium phosphide high electron mobility transistors) and GaAs-mHEMTs (gallium arsenide metamorphic high electron mobility transistors) have shown capability to yield power amplification up to 1.2 THz at gate lengths 30–40 nm [14]. Recently an InP-HEMTs based compact

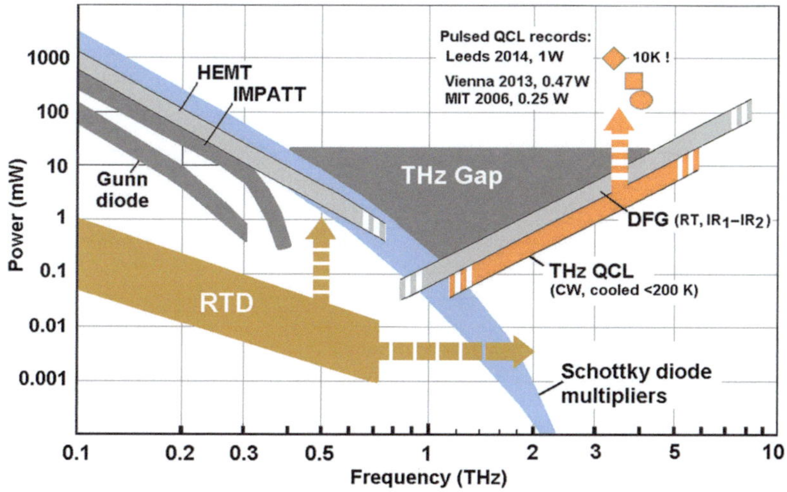

Fig. 20.1 Power versus frequency map of available compact terahertz sources. The "THz gap" indicates the central frequency interval in which is hard to generate any larger radiation power. Note that because QCLs need cryogenic cooling, and both Schottky multipliers and DFGs require significant input power, the compactness of those devices is somewhat questionable (Abbreviations: *QCL* Quantum Cascade Laser, *DFG* Difference Frequency Generation, *RTD* Resonant Tunneling Diode, *HEMT* High Electron Mobility Transistor, *IMPATT* IMPact ionization Avalanche Transit Time diode, *CW* Continuous Wave, *RT* Room Temperature)

amplification module was announced, yielding output power 0.93 mW at 0.85 THz [15]. In contrast to this, the high expectations associated with the graphene transistors have not been confirmed in real experiments due to their low power amplification capability (in comparison with current amplification), leaving the maximum usable frequency even below 0.1 THz [14]. An important universal and popular class of solid state sources are the Schottky diode multipliers (usually doublers and triplers) that may reach in multi-stage cascade configuration the frequencies up to 2.2–3 THz [5, 16]. Compactness of this approach is somewhat limited by the low power conversion efficiency between 0.1 and 1 % of the multiplier stages [5, 16]. Fig. 20.1 also shows the region associated with the resonant tunneling diodes (RTD) on the basis of earlier publications from 1990s [17, 18]. The problem of RTDs has been the low output power due to need for suppressing device capacitance by using very small cross-section area. In recent years remarkable progress has been achieved in moving towards higher frequencies and output powers (see next section).

In conclusion, Fig. 20.1 reveals that the most difficult region for development of THz technologies lies between frequencies 0.5 and 3 THz. If looking from the security applications viewpoint, the characteristic absorption spectra peaks of many important explosives coincide noticeably with the proposed THz gap. To confirm this, Fig. 20.2 shows absorption spectra of several explosive substances [19–21].

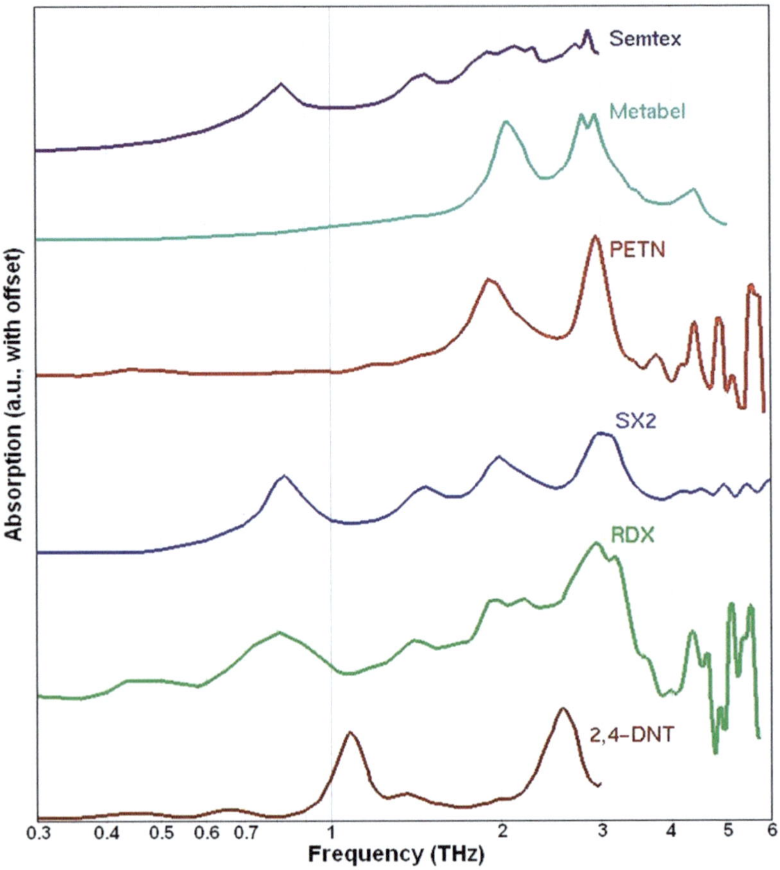

Fig. 20.2 Absorption spectra of some important explosive substances. Note that identification peaks coincide noticeably with the THz gap, and the range below 1 THz cannot be omitted (The data are from Refs. [19–21])

20.2 Development of RTDs Based THz Sources

In the laborious development of THz technologies over the past 50 years the important two steps forward were the realization of the semiconductor heterostructure quantum cascade laser (QCL), first in the mid-infrared range in 1994 [22] and after in the THz-range in 2002 [23]. After these two achievements the number publications under the topics "terahertz" increased over the level of 100 and 400 per year, reaching presently the total number over 2500 per year (Fig. 20.3).

Figure 20.3 demonstrates that actually the QCL research (including both mid-IR and THz ranges) has reached a saturation level that might be explained by approaching the fundamental limits of the QCL operation temperatures and

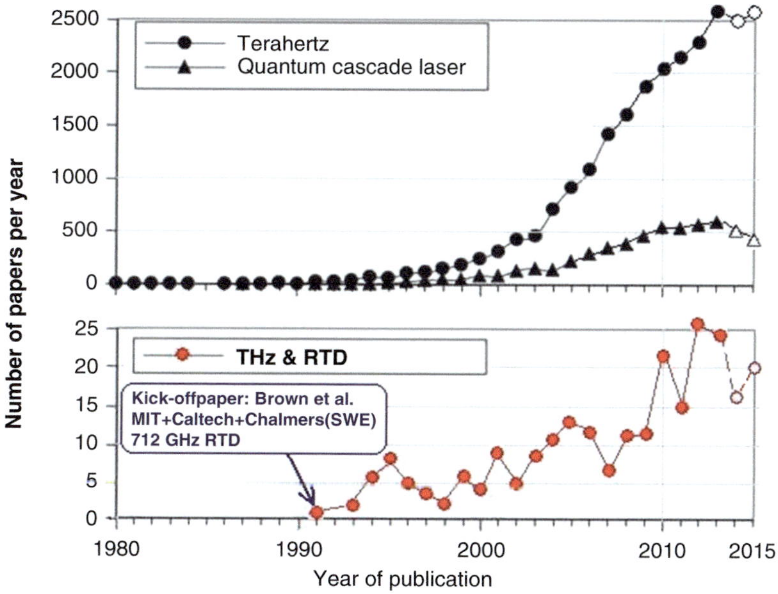

Fig. 20.3 Intensity of terahertz, quantum cascade laser and resonant tunneling diode research (Data from Web of Science)

frequencies. The interest of researchers has thus turned towards other compact THz source solutions. But Fig. 20.3 shows also that RTD research has been much underestimated field as only 1% of all THz publications has been dedicated to those record-fast room temperature electronic devices, which have no problems with the low-power compact realization.

It should be mentioned that although the practical realization of THz RTD sources started in 1991 [17], the idea of using negative differential resistance of biased superlattices for amplification and generation of high frequency oscillations stems from the same Kazarinov and Suris paper from the year 1971, where was proposed the idea of QCLs [24].

Figure 20.4 shows the overall progress of the THz RTD sources development starting from the kick-off work by Brown et al. (MIT and Caltech) in 1991 [17]. -Two-barrier RTDs of Brown group were fabricated on the basis of three different material systems including the InAs/AlSb which has the barrier height of 1350 meV, about the highest possible with A3B5 semiconductors. Although the layer thicknesses were approximately 1.5–2 times larger and the cross section areas even 10–30 times larger, compared to the modern 1.86 and 1.92 THz record devices of Asada group [25, 26], the impressive oscillation frequency 712 GHz was obtained. Brown et al. have also estimated the transit times of tunneling barriers and depletion layer to be below 100 f. and the theoretical maximum frequency of their design 1.24 THz.

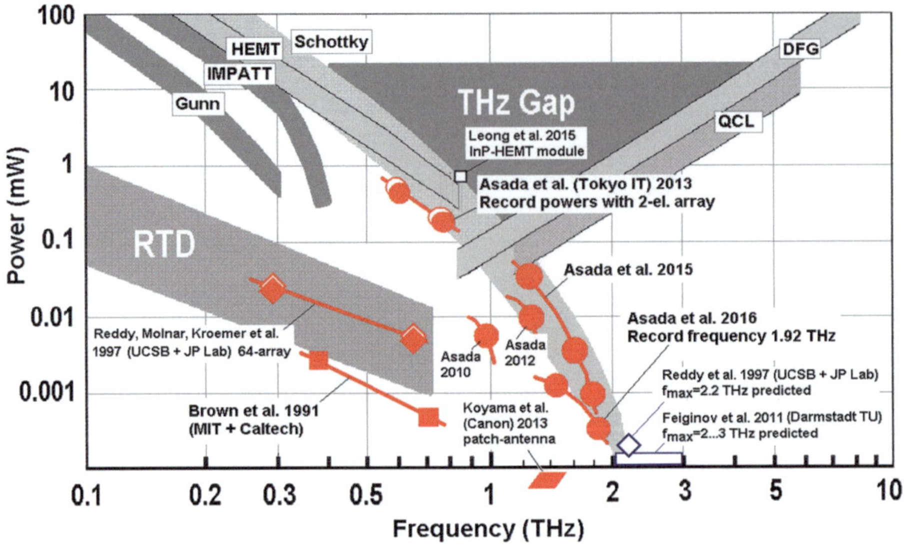

Fig. 20.4 Progress in the development of resonant tunneling diodes based THz sources on the power versus frequency map (See explanations in text)

The next important step forward was made by Reddy, Molnar, Kroemer et al. in 1997 (University of California in Santa Barbara and Jet Propulsion Lab.) by designing a 64-element RTD-array [18] to overcome the low output power problem. It should be mentioned the negative differential conductance makes the parallel (and series) connection of multiple RTDs difficult, and a special inductive separation of DC bias and AC oscillations must be added. Reddy et al. used InGaAs/AlAs material system, similar to later work of Asada group [25–28]. The 64-element array yielded approximately tenfold improvement in power compared to previous work [17] but this remained approximately 100 times below the power obtained later in 2013 by Asada et al. with just 2-element RTD arrays [27]. Reddy et al. predicted the theoretical frequency limit 2.2 THz of their design [18].

Further progress in development of THz RTD oscillators is dominated by the Masahiro Asada group from Tokyo Institute of Technology [25–29, 31]. The group started THz RTD research in 2005 [28] and has been focused on GaInAs/AlAs material system (barrier height ~800 meV) and optimization of the compact integrated slot antenna design, shown schematically in Fig. 20.5 below.

With the basic design sketched in Fig. 20.5, Asada group showed in 2010 the record frequency 1.04 THz with noticeable output power 7 μW [29]. Very thin 1.2 nm AlAs barrier layers and 3-step graded InGaAs emitter layer were used here, and slot antenna length was 20 μm.

In 2011 the oscillation frequency record 1.11 THz was taken by the Feiginov group from Darmstadt Technical University [30], but output power of their design remained rather low, at 0.13 μW. However, they predicted that it is possible to reach frequencies 2–3 THz after optimization of their oscillator design [30]. The layout of

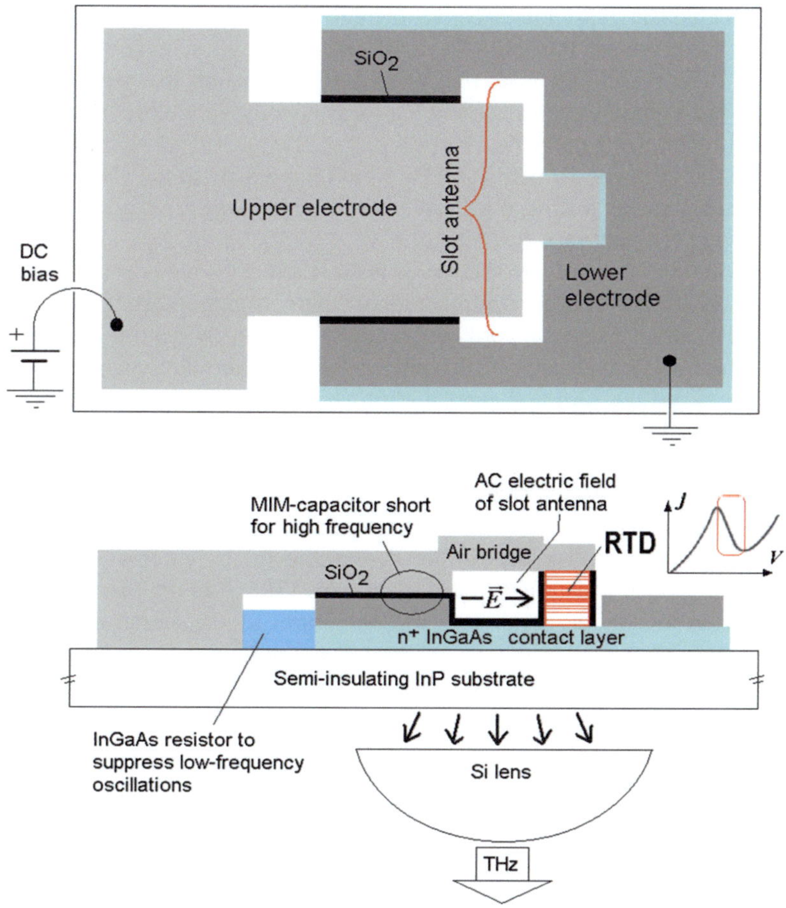

Fig. 20.5 Schematic presentation of the RTD oscillator design with integrated slot antenna approach developed by M. Asada group from Tokyo Institute of Technology [25–29, 31].

the Feiginov group design was very different from Asada approach shown in Fig. 20.5 and used a larger area but still below 1 mm^2.

In 2012 Asada group shifted frequency record to 1.31 THz, with impressive output power 10 μW using narrow 1 nm barriers and 3.9 nm central quantum well [31]. The mesa area was reduced to 0.33 μm^2 while keeping the antenna slot length at 20 μm. It was predicted that for larger output power the asymmetric feeding position of RTD inside the antenna slot should be used.

Following this, using offset-fed structure together with longer antenna length of 130 μW together with careful matching of the antenna and RTD impedances, and also using the 2-element RTD arrays, the record high output powers 0.61 and 0.18 mW were obtained in 2013 for sub-THz frequencies 0.62 and 0.81 THz, respectively [27]. At that the 2-element array gave very significant power increase by factor of 1.8. The 4-element arrays were also fabricated but then mutual injection locking mechanism was not effective enough to yield the necessary match of oscillation frequencies. It should be mentioned that the achieved power range, close to 1 mW at 0.5 THz and 10 μW at 1.3 THz, meant that the performance area of RTDs on the power-frequency map was already very close to the Schottky diode multipliers, see Fig. 20.4.

In 2014 the frequency record 1.4 THz went temporarily to the Canon team [32] that was using an interesting patch antenna design. However their output power remained very low, below a few tens of nW.

In 2015 and in January 2016 the Asada team reached the new record frequencies 1.86 and 1.92 THz by further optimization of their structure sketched in Fig. 20.5 [25, 26]. To achieve these record frequencies, the narrow central well width of 2.5 nm, small RTD mesa area of 0.1 μm^2, relatively short slot antenna of 12 μm, wide 6 μm air bridge [26] and careful elimination of air bridge skin effect losses [26] were used. The values 20–35 fs and 80 fs were estimated for the intrinsic delay time (RTD layers) and external delay time (antenna capacitances and resistances), respectively [25]. Very short internal delay time showed great potential of improving of the frequency performance. It was estimated that to reach the oscillation frequencies beyond 2 THz the further decreasing of the external delay time is needed, using for this also the thicker electrodes [26]. Authors mention that they have not yet fabricated the small mesa area structures for the shorter antenna length of 9 μm for which the extrapolated curves indicate possibility to reach frequencies beyond 2.2–2.4 THz [26]. At lower frequencies 1.3–1.5 THz the obtained power values [25] were already fully comparable with those of Schottky diode multipliers, see Fig. 20.4.

20.3 Conclusion

In this chapter we analyzed the remarkable progress that has occurred in the development of resonant tunneling diodes as genuine compact sub-millimeter size range and room temperature THz sources. Due to continuous search for ever better solutions over the last 10 years the achieved power range of RTDs has been increased more than 2 orders of magnitude and the frequency range has been shifted from initial 0.7 THz to nearly 2 THz. Thus the performance of RTDs has now become already quite comparable to the performance of the popular Schottky diode based frequency multiplication systems.

Acknowledgments The authors acknowledge support from COST action MP1204 – TERA-MIR Radiation: Materials, Generation, Detection and Applications and from COST action BM1205 – European Network for Skin Cancer Detection using Laser Imaging.

References

1. D. L. Woolard et al., Terahertz frequency sensing and imaging: A time of reckoning future applications? IEEE Proceedings, 93, 1722–1743 (2005)
2. D. Saeedkia (ed), Handbook of terahertz technology for imaging, sensing and communication. Woodhead Publishing series in electronic and optical materials, 34 (Woodhead Publishing Ltd., Oxford/Cambridge/Philadelphia/New Delhi, 688 p., 2013)
3. J. Faist, Quantum Cascade Lasers (Oxford Univ. Press, 320 p, 2013)
4. D. P. Burghoff, Broadband terahertz photonics. PhD thesis (Massachusetts Institute of Technology, 2014)
5. D. L. Woolard, W. L. Loerop, and M. S. Shur (eds), Terahertz sensing technology. Volume 1: Electronic devices and advanced systems. (World Scientific Publishing, Singapore, 2003)
6. D. J. Paul, The progress towards terahertz quantum cascade lasers on silicon substrates. Laser Photonics Rev., 4, 610–632 (2010)
7. R. Reeder, E. Velmre, and A. Udal, Advances in terahertz technology with emphasis on quantum cascade lasers. Electronics and Electrical Engineering, 8(104), 47–50 (2010)
8. M. A. Belkin, F. Capasso, A. Belyanin et al., Terahertz quantum-cascade-laser source based on intracavity difference-frequency generation. Nature Photonics, 1, 288–292 (2007)
9. Q. Lu et al., Room temperature continuous wave, monolithic tunable THz sources based on highly efficient mid-infrared quantum cascade lasers. Sci. Rep., 6, 23595 (2016). doi: 10.1038/srep23595
10. S. Kumar, Recent progress in terahertz quantum cascade lasers. IEEE J. Sel. Topics in Quantum Electr., 17, 38–47 (2011)
11. M. Amanti, G. Scalari, R. Terazzi et al., Bound-to-continuum terahertz quantum cascade laser with a single-quantum-well phonon extraction/injection stage. New J. of Physics, 11, 125022 (2009)
12. L. H. Li, J. X. Zhu, L. Chen, A. G. Davies, and E. H. Linfield, The MBE growth and optimization of high performance terahertz frequency quantum cascade lasers. Optics Express, 23, 2720–2729 (2015)
13. L. H. Li, J. X. Zhu, L. Chen, J. Freeman, P. Dean, A. Valavanis, A. G. Davies, and E. H. Linfield, Terahertz quantum cascade lasers with >1 W output powers. Electronics Lett., 50, 309–311 (2014)
14. F. Schwierz, Graphene transistors: status, prospects, and problems. IEEE Proceedings, 101, 1567–1584 (2013)
15. K. M. Leong et al., A 0.85 THz low noise amplifier using InP HEMT transistors. IEEE Microw. Wireless Compon. Lett., 25, 395–399 (2015)
16. J. L. Hesler and T. Crowe, High power solid state terahertz sources. SPIE Newsroom homepage (2015). doi: 10.1117/2.1201506.005859
17. E. R. Brown, J. R Söderström, C. D. Parker, L. J. Mahoney, K. M. Molvar, and T. C. McGill, Oscillations up to 712 GHz in InAs/AlSb resonant tunneling diodes. Appl. Phys. Lett., 58, 2291–2293 (1991)
18. M. Reddy, S. C. Martin, A. C. Molnar et al., Monolithic Schottky-collector resonant tunnel diode oscillator arrays to 650 GHz. IEEE Electron. Dev. Lett., 18, 218–221 (1997)
19. A. G. Davies, A. D. Burnett, W. Fan, E. H. Linfield, J. E. Cunningham, Terahertz spectroscopy of explosives and drugs. Materials Today, 11, 18–26 (2008)
20. J. Chen et al., Absorption coefficients of selected explosives and related compounds in the range of 0.1–2.8 THz. Optics Express, 15, 12060–12067 (2007)
21. M. R. Leahy-Hoppa, M. J. Fitch, X. Zheng et al., Wideband terahertz spectroscopy of explosives. Chem. Phys. Lett, 434, 227–230 (2007)
22. J. Faist et al., Quantum cascade laser. Science, 264, 553–556 (1994)
23. R. Köler et al., Terahertz semiconductor-heterostructure laser. Nature, 417, 156–159 (2002)
24. R. F. Kazarinov and R. A. Suris, Possibility of amplification of the electromagnetic waves in a semiconductor with a superlattice. Soviet Physics – Semiconductors, 5, 707–709 (1971)

25. H. Kanaya, T. Maekawa, S. Suzuki, and M. Asada, Structure dependence of oscillation characteristics of resonant-tunneling-diode terahertz oscillators associated with intrinsic and extrinsic delay times. Jap. J. Appl. Phys., 54, 094103 (2015)
26. T. Maekawa, H. Kanaya, S. Suzuki, and M. Asada, Oscillation up to 1.92 THz in resonant tunneling diode by reduced conduction loss. Appl. Phys. Express, 9, 024101 (2016)
27. S. Suzuki, M. Shiraishi, H. Shibayama, and M. Asada, High-power operation of terahertz oscillators with resonant tunneling diodes using impedance-matched antennas and array configuration. IEEE J. Selected Topics in Quantum Electronics, 19, 8500108 (2013)
28. N. Orihashi, S. Hattori, S. Suzuki, and M. Asada, Voltage-controlled sub-terahertz oscillation of resonant tunneling diode integrated with slot antenna. Electronics Lett., 41, 872–873 (2005)
29. S. Suzuki, M. Asada, A. Teranishi et al., Fundamental oscillation of resonant tunneling diodes above 1 THz at room temperature. Appl. Phys. Lett., 97, 241102 (2010)
30. M. Feiginov, C. Sydlo, O. Cojocari, and P. Meissner, Resonant-tunneling-diode oscillators operating at frequencies above 1.1 THz. Appl. Phys. Lett., 99, 233506 (2011)
31. H. Kanaya, H. Shibayama, S. Suzuki, and M. Asada, Fundamental oscillations up to 1.31 THz in thin-well resonant tunneling diodes. Paper presented at the 2012 Int. Conf. on InP and related materials (IPRM), 106–109 (2012). doi: 10.1109/ICIPRM.2012.6403331
32. Y. Koyama, R. Sekiguchi, and T. Ouchi, Oscillations up to 1.4 THz from resonant-tunneling-diode-based oscillators with integrated patch antennas. Appl. Phys. Express, 6, 064102 (2013)

Chapter 21
Investigating Glow Discharge Detectors as a Millimeter-Wave/Terahertz Radiation Detection Tool

N. Alasgarzade, M.A. Nebioğlu, T. Takan, I.U. Uzun-Kaymak, A.B. Sahin, and H. Altan

Abstract Compared to other forms of electromagnetic radiation, Terahertz (THz) radiation is considered safe for imaging and detection purposes due to their non-invasive and non-destructive nature. Novel techniques and methods are constantly being evaluated for efficient detection and measurements of THz waves, yet their commercial success is limited due to cost and overall complexity of these systems. Commercially available Glow Discharge Detectors (GDDs) are proven to detect microwave and higher frequency radiation. Although, the plasma generated inside the GDDs enables the detection of the EM radiation, the interaction mechanism is not fully understood. In this study, we investigate various types of GDDs that are commercially available, and characterize their overall behaviour in response to both CW and modulated EM radiation. In addition, the response from various GDDs to the mm-wave/THz radiation are surveyed for the range of 260–380 GHz. Measurements show that response signal at certain frequencies within the range are attenuated suggesting a resonance frequency possibly based on the electrode structure.

Keywords THz radiation • Mm-wave radiation • Glow discharge detector

21.1 Introduction

Millimetre wave and terahertz (THz) regions of the electromagnetic spectrum which correspond to 30–300 GHz and 0.3–10 THz bands respectively, promise many novel applications in various fields owing to their unique properties. THz and

N. Alasgarzade • M.A. Nebioğlu • T. Takan • I.U. Uzun-Kaymak • H. Altan (✉)
Department of Physics, Middle East Technical University, Ankara, Turkey
e-mail: haltan@metu.edu.tr

A.B. Sahin
Department of Electrical and Electronics Engineering, Yildirim Beyazit University, Ankara, Turkey

M.F. Pereira, O. Shulika (eds.), *THz for CBRN and Explosives Detection and Diagnosis*, NATO Science for Peace and Security Series B: Physics and Biophysics,
DOI 10.1007/978-94-024-1093-8_21

mm-waves are capable of penetrating most non-metallic and non-polar materials. This property alone can be utilized to identify concealed substances and weapons through clothing and other barriers such as walls and soil. In addition, hazardous gases, illegal drugs and explosives are proven to have unique signatures in THz spectra, this regime of electromagnetic spectrum has great potential for security screening [1, 2]. Despite the promise of novel applications, detection of mm and THz waves at room temperature is a challenging task. There are various uncooled detectors for direct detection of mm-wave/THz radiation, namely Golay cells, photoconductive antennae, pyroelectric sensors, piezoelectric resonant structures, Si FET type devices, Schottky diodes, and bolometer based detectors. These detectors have noise equivalent power (NEP) values that range from 10^{-8} to 10^{-12} W/Hz$^{1/2}$. Golay cells, and pyroelectric sensors typically have a broad spectral response covering 0.1–10 THz. Operating at low modulation frequencies and having long response times are disadvantages of Golay cells, whereas pyroelectric sensors can operate at high modulation frequencies above a few kHz. Nonetheless, these detectors share a common drawback of being difficult to implement into an array to be used as an imaging sensor [3].

Recent studies show that a miniature neon indicator lamp, also known as the Glow Discharge Detector (GDD) is, in fact, a low-cost alternative of such detectors, since GDDs have a fast response time, broad spectral response, low NEP and room temperature applicability for both detection and imaging purposes [4]. Detection of mm and THz waves using GDDs can be explained by electromagnetic wave-plasma interaction phenomena. Working principal of a neon indicator lamp involves a DC capacitively coupled gas discharge. The gas used can be any inert gas, however commercially available lamps are often filled with Neon penning mixtures, to enhance ionization. The discharge process starts with applying a DC bias voltage between the two electrodes. The electrode separation is measured to be about 1 mm. In addition, the primary electrode material in certain types of commercially available lamps is nickel, which is then coated with materials to reduce the breakdown voltage. Since the plasma frequency of ionized gas inside the tube is about a few GHz [5], the plasma becomes transparent to incident radiation of higher frequencies ($\omega > \omega_p$). Strong electric field between electrodes leads to move free electrons from cathode to anode. The incident electromagnetic wave is absorbed by free electrons and this process cause slight changes in current between electrodes. Enhanced ionization and enhanced diffusion mechanisms play key role in detecting incident electromagnetic wave with plasma. Change in current can be positive or negative depending on which detection mechanism is dominant. Detected signal becomes positive when electric field of incident radiation enhances ionization rate more than diffusion. Change in current is given by

$$\Delta I_d = (\Delta \nu_i) \times n + \nu_i \times (\Delta n) - (\Delta D) \times \nabla^2 n - D \times \nabla^2 (\Delta n) \qquad (21.1)$$

where ν_i is ionization collision frequency, D is diffusion coefficient and n is charge-carrier volume density. Since main carrier in process is electron rather than ion, n is the electron number density [6].

Total electron energy depends dc electric field existing between the electrodes. Generated electric field has its maximum at cathode dark space and minima at negative glow and faraday space regions of discharge [7]. Ionization processes mostly take place at Cathode dark space where total electron energy has its maximum. Electrons at faraday dark space and negative glow regions begin to oscillate due to absorption of incident radiation and diffuse out these regions toward to walls of neon lamp. Both mechanisms appear in the detection process of the radiation at same time, but one of them becomes dominant. Studies have shown that dominant mechanism in detecting millimetre wave by using GDD is enhanced cascade ionization [8]. Small increase in kinetic energy of free electrons allows them to participate in ionization and excitation processes by colliding with atoms. If the energy of the accelerated electrons is equal or greater than the ionization energy of atoms inside the lamp, they ionize atoms by removing weakly bound electron of the atom hence new electrons are generated. After the collision, some electrons lose their energy, however some gain energy by absorbing THz radiation so they can participate in ionization process again. Moreover, THz photons can also contribute to the ionization process by exciting neon atoms. New electrons are constantly generated after several ionization processes and this effect can be observed as an increase in current. Change in current as a function of time is formulated as,

$$\Delta I_d(t) = \frac{G \times e^2 \times V \times n}{V_i \times m}\left(\frac{\tau}{\tau_i}\right) \times \eta_0 \times P_D \times \left(\frac{\nu}{\nu^2 + \omega^2}\right) \times \left[1 - \exp\left(\frac{-t}{\tau}\right)\right] \tag{21.2}$$

$$G \approx \frac{\exp(2 \times \nu_i \times t_d)}{2 \times \nu_i \times t_d} \tag{21.3}$$

where V is discharge volume, V_i is the effective ionization potential of the gas inside lamp, t_d is average electron drift time to the anode, τ is detection time constant, τ_i average time interval between two RF electron ionizing collisions, P_D is average power density of incident electromagnetic wave and G is internal signal gain. Theoretically maximum responsivity of GDD obtained when elastic electron-neutral frequency (ν) is equal to frequency of radiation (ω). Moreover, the response of the detector linearly depends on the power of incident radiation [4].

21.2 Experimental Setup and Results

In order to study response of GDDs at various modulation frequencies of incident mm-wave, an experimental set-up shown in Fig. 21.1 is constructed. Two types of neon lamps namely N523 and N527 commercially obtained from International Light Technology (Peabody, MA) are discussed in this paper since they provide the best detected signal levels among various types of GDDs. These neon lamps

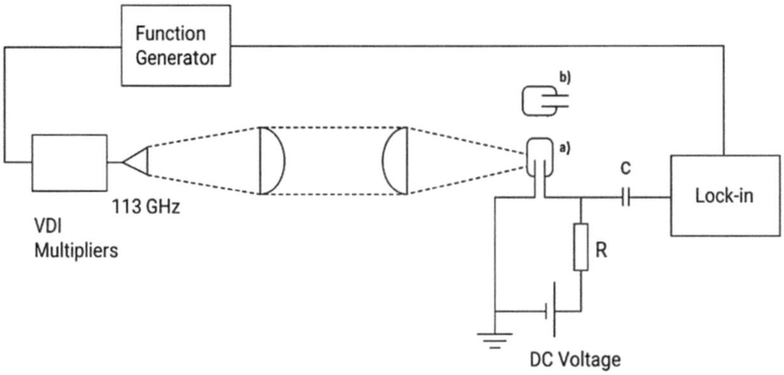

Fig. 21.1 Experimental set-up used to measure response of GDD illuminated with 113 GHz radiation in case of two different orientations of GDD: (**a**) leads are perpendicular and (**b**) leads are parallel to the direction of incident beam (on-axis)

have phosphorus coating which may cause an increase in detected signal [8]. The electrode gap of GDDs is about 1 mm and casing diameters are 6 mm and 4.2 mm for N523 and N527 respectively. The transmitter (Virginia Diodes Inc., 110–120 GHz Amplified Multiplier Chain) with a 113 GHz output frequency is used as a mm-wave source. The transmitter was in a modified configuration where the Dielectric Resonator Oscillator (DRO) source can be modulated up to tens of MHz by use of a mixer driven by a function generator. Modulated beam is focused on the electrode gap by using a set of Teflon lenses (50 mm diameter, F#2). The beam diameter is approximately 8 mm. By using Golay Cell (Tydex GCP-1), the beam power is measured to be on the order of 20 mW at low modulation frequencies. In bias circuit of GDD, a resistor is used to limit the discharge current on the order of milliamps, which is sufficient to force lamp to operate in the abnormal glow where most gas atoms inside tube are in the excited state [9]. A capacitor is used to by-pass the DC signal so that only the fluctuating signal generated from current change caused by the incident radiation is measured. Compared to other uncooled room temperature mm-wave/THz detectors, GDD can operate at higher modulation frequencies. The response of GDDs at modulation frequencies up to 90 kHz is analyzed by using a lock-in amplifier (SRS 530). The maximum frequency response of the lock-in is limited to 100 kHz. In order to investigate modulation frequency dependence of these lamps, we applied 130 V DC between the electrodes. Response of GDDs to different modulation frequencies is measured for frequencies between 1 and 90 kHz as shown in Fig. 21.2. Response of both GDDs increase with the increase in modulation frequency in both orientations: leads are perpendicular and leads are parallel to the direction of incident beam (on-axis). The N527 shows better performance than the N523 in both orientations. It is interesting that the detected signal for N527 is much higher for on-axis configuration than perpendicular. This can be interpreted such that the incident radiation interacts more with native plasma inside tube along the on-axis orientation

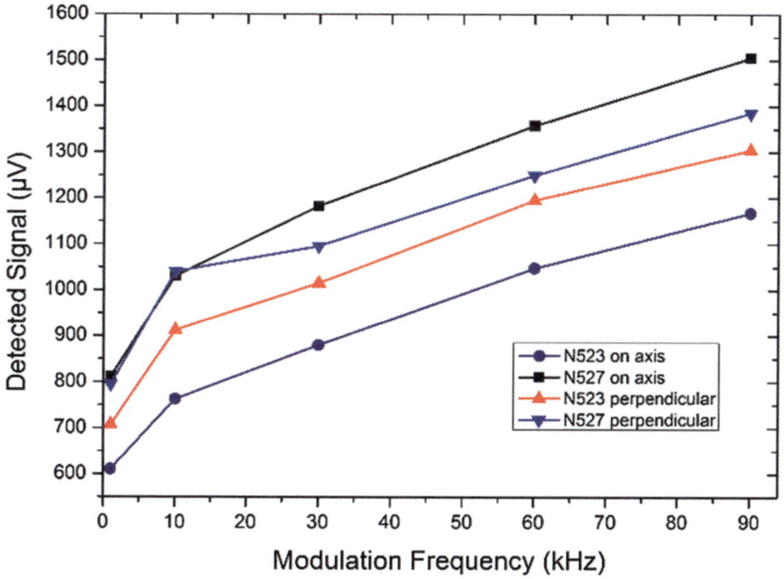

Fig. 21.2 Comparison of the detected signal for the two best neon indicator lamps (N523 and N527)

rather than the perpendicular orientation, which is consistent with the earlier finding listed in Reference [10]. The observed signal dependence on the orientation can also be explained by the fact that this particular GDD casing has a bulbous tip which may scatter the incident radiation for the on-axis configuration. In addition, it is observed that the GDD structure is highly sensitive to the orientation of the polarization of the incident radiation with a behavior that closely follows the Malus' law [11].

The GDD has a dipole antenna-like structure which is thought to aid in the detection of the mm-wave radiation [12]. Since the electrode spacing is about 1 mm, some frequencies may be resonant with the dimensions of the structure of the device. Using another experimental configuration (see Fig. 21.3), the frequency dependence of GDD- THz interaction is analysed thoroughly in the 260–380 GHz frequency range using a VDI multiplied source (WR2.8AMC) driven by a tuneable Yttrium Iron Garnett (YIG) oscillator. The GDD is placed at the focus of a 50 mm diameter Teflon lens (F#2) and the radiation passing through the leads are collected in both scenarios: GDD lamp on and off. The fine tunability of the system allows for measurements with 0.5 GHz precision. The power of transmitted radiation is measured by using a Golay cell. Glow discharge process starts with applying 110 V between the electrodes of neon lamp, N523 in particular. The applied dc voltages are 110, 130 and 145 V, corresponding to discharge currents on the orders of milliamps. In Fig. 21.4, the transmission represents the ratio of power transmitted when the GDD is on to that when it is off. The absorption of the incident THz radiation by the GDD increases with an increase of bias current at specific

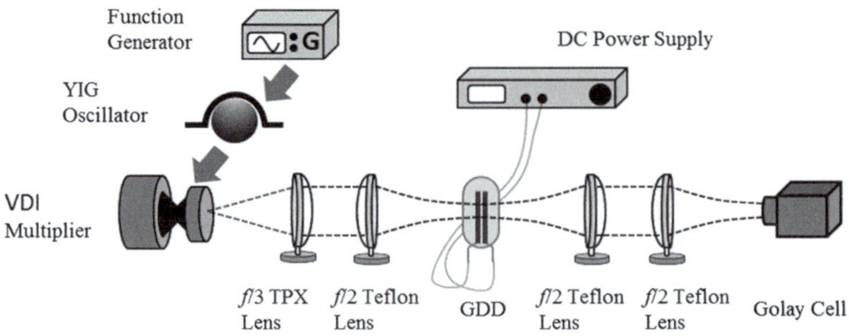

Fig. 21.3 Schematic representation of the mm-wave/THz measurement system, working in 260–380 GHz region

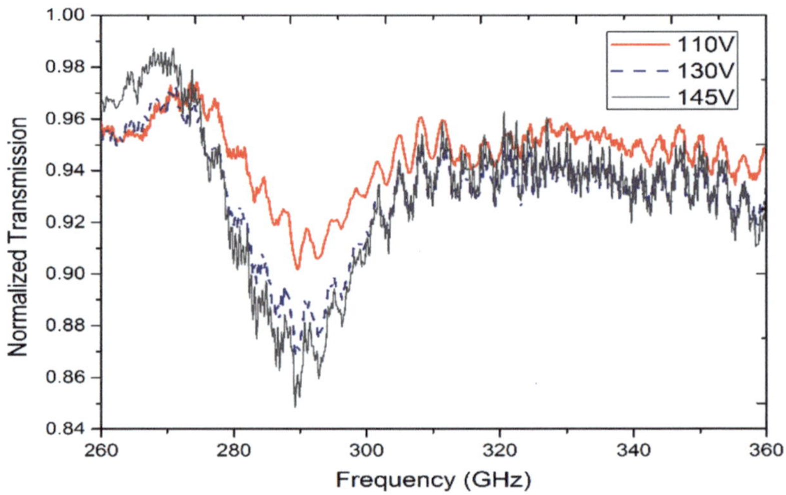

Fig. 21.4 Transmission of the mm-wave/THz radiation through the GDD (N523) at different DC bias voltages

frequency. These measurements are compared to previous studies where the GDD was turned on and placed at the focus of a broadband THz beam, the THz transmission was attenuated at certain frequencies [13].

In addition to these experimental results, using CST Microwave Studio (CST), a commercial electromagnetic simulation software, the transmission characteristics of plasma–THz interaction is studied numerically. The interaction of the THz-wave with the plasma can be described using the Drude model. According to the Drude model, the dominant charge carriers are electrons rather than ions since electrons' speed are much higher than ions. For glow discharge plasmas, the electron density is often considered in the range of $10^{14}\ \mathrm{m}^{-3} \leq n_e \leq 10^{16}\ \mathrm{m}^{-3}$. For this density range, the electron plasma frequency is within $0.09\ \mathrm{GHz} \leq f_{pe} \leq 0.9\ \mathrm{GHz}$.

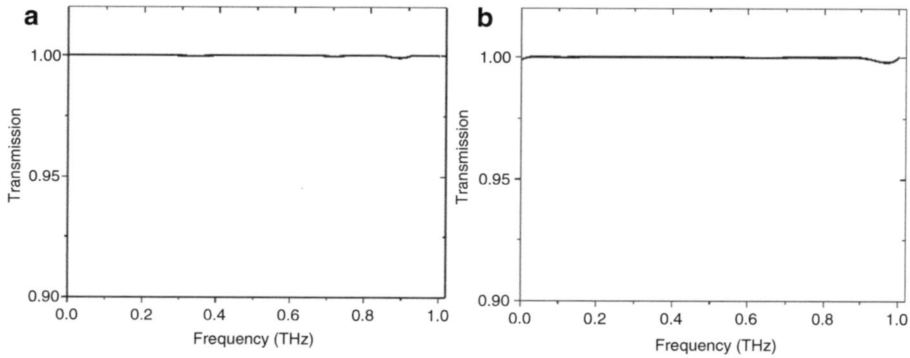

Fig. 21.5 Simulated transmission spectrum of the THz-wave through plasma with different electron plasma frequencies. (**a**) Plasma frequency is 900 MHz and electron neutral collision frequency is 16.7 GHz. (**b**) Plasma frequency is 1 GHz and electron neutral collision frequency is 55.7 GHz

Using CST, various possibilities are investigated. First, assuming a steady glow with Maxwellian distribution the electron collision rate is expected to be much less than the plasma frequency, i.e. on the order of 100 kHz to few MHz. CST simulations show that these types of collisions among charged particles do not change the transmission of the THz in general, hence the observed phenomena cannot be explained using a steady glow.

On the other hand, if one assumes that in the abnormal glow region, the electrons are highly energetic, and by assuming partial atmospheric pressure inside the GDD lamp is within the range of 30–100 torr, and the gas inside tube is pure neon, one can estimate electron neutral collision frequency for intended pressure by using $\nu \approx (3.5 \times 10^9)p$ [14, 15]. Here, the electron neutral collision frequency is found as 16.7 GHz and 55.7 GHz for two different gas pressures 30 and 100 torr, respectively. The transmission spectrum of THz-wave modeled using CST is shown in Fig. 21.5, for different collision frequencies. In case of high collision rate, there are some miniscule ripples in the transmission, although these transmission numbers are much smaller compared to the observed transmission dip. Therefore, we suspect that the observed dip is probably due to the electrode configuration and the plasma is transparent to incident electromagnetic wave within the range of 0.001–1 THz.

21.3 Conclusion

Two commercially available neon indicator lamps with phosphorus coatings are characterized in the mm-wave/sub-mm-wave/THz region of the electromagnetic spectrum. It is found that the response of GDD increases with increasing modulation frequency. The N527 provides the best performance as a mm-wave/THz detector. The response of the GDD is also characterized near the low-THz end of the spectrum to understand any resonant effects between the structure and the incident wavelength. Results for the GDDs with the best signal response show that the transmission of the sub-mm-wave radiation decreases at certain frequencies. Contrast to the experimental results, numerical simulations show no absorption at specified frequencies. The attenuation of signal is suspected to be a signature of the electrode geometry of GDD. Although GDDs have potential to be a cost-effective mm/sub-mm-wave detector, further studies including the effects of electrode spacing, gas type and encasing are needed to pinpoint the exact characterization of the detector.

Acknowledgments This research is sponsored by NATO's Public Diplomacy Emerging Security Challenges Division in the framework of "Science for Peace "the Science for Peace and Security Programme under grant MD.SFPP 984775 and under The Scientific and Technical Research Council of Turkey (TUBITAK) grant # 113F321.

References

1. M. Tonouchi, Cutting-edge terahertz technology, Nature Photonics, 1, 97–105 (2007).doi:10.1038/nphoton.2007.3
2. J. F. Federici, B. Schulkin, F. Huang, D. Gary, R. Barat, F. Oliveira, D. Zimdars, THz imaging and sensing for security applications—explosives, weapons and drugs, Semicond. Sci. Technol. 20, S266–S280 (2005). doi:10.1088/0268-1242/20/7/018
3. A. Rogalski, F. Sizov, Terahertz detectors and focal plane arrays, Opto-electronics review, 19, 346–404 (2011). doi: 10.2478/s11772-011-0033-3
4. A. Abramovich, N. S. Kopeika, D. Rozban, E. Farber, Inexpensive detector for terahertz imaging, Appl. Opt., 46, 7207–7211 (2007). doi: 10.1364/AO.46.007207
5. N.S. Kopeika, Glow discharge detection of long wavelength electromagnetic radiation: Cascade ionization process internal signal gain and temporal and spectral response properties, IEEE Trans. Plasma Sci., 6, 139–157 (1978). doi: 10.1109/TPS.1978.4317104
6. N.S. Kopeika , On the mechanism of glow discharge detection of microwave and millimeter wave radiation, Proc. IEEE 63, 981–982 (1975).doi:10.1109/PROC.1975.9864
7. B. Chapman, Glow Discharge Processes: Sputtering and Plasma Etching, Wiley-Interscience publication (Wiley, New York, 1980.) p. 78–79
8. D. Rozban, N.S. Kopeika, A. Abramovich, E. Farber, Terahertz detection mechanism of inexpensive sensitive glow discharge detectors, J.Appl. Phys., 103, 093306-1–093306-4 (2008). doi: 10.1063/1.2917386
9. K. Çınar, H. Altan, A.B. Şahin, THz transmission and detection through glow discharge detectors, Proc. SPIE 8188, 81880A–8188A-7 (2011). doi:10.1117/12.2017967
10. A. Abramovich, N.S. Kopeika, D. Rozban, THz polarization effects on detection responsivity of Glow Discharge Detectors (GDD), IEEE Sensors J., 9, 1181–1184 (2009). doi:10.1109/JSEN.2009.2027415
11. T. Takan, N. Alasgarzade, I. U. Uzun-Kaymak, A. B. Sahin, and H. Altan, Detection of far-infrared radiation using glow discharge detectors, Optical and Quantum Electronics, 48, 292 (2016). doi: 10.1007/s11082-016-0569-z
12. A.A. Aharon, D. Rozban, N.Banay, A.Abramovich, N.S. Kopeika, A. Levanon, Polarization effects on heterodyne detection and imaging using Glow Discharge Detector at millimeter wavelengths, Proc. SPIE 9078, 90780F (2014). doi: 10.1117/12.2053175

13. K. Çınar, H.M. Bozacı, H.Altan, Characterization of a glow discharge detector with terahertz time domain spectroscopy, IEEE Sens. J., 13, 2643–2647 (2013). doi: 10.1109/JSEN.2013.2258666
14. N. H. Farhat, Optimization of millimeter-wave glow-discharge detectors, in *Proc. IEEE*, 62, 2, 279–281, (1974). doi: 10.1109/PROC.1974.9416
15. A.D. MacDonald, Microwave Breakdown in Gases, ed.by J. Wiley and Sons (New York, 1966) p. 24–25

Chapter 22
The Linewidth Enhancement Factor of Dilute Nitride Intersubband Lasers Without Inversion

Mauro Fernandes Pereira

Abstract The linewidth enhancement (α factor) due to fluctuations in the refractive index induced by carrier fluctuations of intersubband lasers was initially expected to be zero. However, values ranging from −0.5 to 3 have been found experimentally. For laser without inversion conditions, which have potential for the high temperature operation of terahertz lasers, α is found to be larger, but still at the same order of magnitude of conventional inverted medium lasers and this paper further investigates this case for dilute nitride quantum well structures.

Keywords Linewidth enhancement factor • Quantum cascade lasers • Dilute nitride quantum wells • Intersubband optics • TERA-MIR radiation

22.1 Introduction

The linewidth of a conventional laser is due to fluctuations in the laser field induced to spontaneous emission and modelled by the Schawlow-Townes formula [1], $\Delta\nu_{ST} = A/n_{ss}$, where A is the spontaneous emission coefficient and n_{ss} is the number of photons in the lasing mode at steady state. In addition to that, in a semiconductor laser there is a contribution arising from fluctuations in the refractive index induced by carrier density fluctuations. The later are quantitatively characterized by the linewidth enhancement or α factor $\Delta\nu = (1 + \alpha^2)\ \Delta\nu_{ST}$ [2, 3].

Intersubband emitters such as Quantum Cascade Lasers (QCLs) made with GaAs and $Ga_{1-x}Al_xAs$ for well barrier layers usually have highly symmetric Lorentzian-like gain profiles, which led to the expectation of zero α factor. Surprisingly, a relatively large spread of values has been found measured, with sub-threshold α factor from −0.5 to 0.5 [4] and above-threshold values (at room temperature) between 0.2 and 3 [5, 6]. For THz QCLs, α between 0.2 and 0.5 has

M.F. Pereira (✉)
Materials and Engineering Research Institute, Sheffield Hallam University, Sheffield S1 1WB, UK
e-mail: M.Pereira@shu.ac.uk

M.F. Pereira, O. Shulika (eds.), *THz for CBRN and Explosives Detection and Diagnosis*, NATO Science for Peace and Security Series B: Physics and Biophysics,
DOI 10.1007/978-94-024-1093-8_22

been found experimentally [7]. A recent theoretical investigation [8] resolved this apparent controversy by demonstrating that counter rotating terms, usually ignored in simulations are the actual fundamental origin of nonzero α at peak gain even without inclusion of nonparabolicity and manybody effects, which are however needed to explain negative values. For laser without inversion conditions, significant as a potential solution to raise the operating temperature of terahertz lasers, α is found to be larger, but still within the order of magnitude of conventional inverted medium lasers. Deeper understanding and control of the linewidth of intersubband emitters is crucial for further spectroscopic applications in the THz and Mid Infrared (TERA-MIR) range of the electromagnetic spectrum [9, 10].

This paper extends and complements the study in Ref. [8] by giving more details of the evolution of the real and imaginary parts of the intersubband optical susceptibility to the limits of gain without inversion and graphically demonstrating the role of rapid peak gain evolution versus small changes in the α factor structure. It is organized as follows: the main equations are outlined. Numerical results are presented, followed by a brief summary.

22.2 Mathematical Approach and Model Equations

The numerical results shown next were obtained with Nonequilibrium Green's Functions (NEGF), which deliver predictive simulations for intersubband transport and optics [8–16].

The α factor, α (E) and gain g(E) spectra as a function of photon energy E stem from the optical susceptibility χ(E) [8]. Note that CGS units are used here.

$$\alpha(E) = \frac{\delta Re\{\chi(E)\}}{\delta Im\{\chi(E)\}}, \quad g(E) = -\frac{4\pi E}{c\hbar n_b}\mathrm{Im}\{\chi(E)\} \tag{22.1}$$

The background refractive index and the speed of light are given respectively by n_b and c. The optical susceptibility is directly obtained in linear response to a probe field from the carriers Green's function (GF). Considering the set of quantum numbers {μ **k**} and explicitly including the interaction of the probe electromagnetic field with matter, and after Fourier transform to frequency (or equivalently Energy) and k-space we obtain an integro-differential equation for the matrix elements and the total optical susceptibility [8]

$$\chi(E) = 2/\Omega \sum_{\mu,\,\nu\neq\mu,\,\boldsymbol{k}} \wp_{\mu\nu\boldsymbol{k}}\, \chi_{\nu\mu}(\boldsymbol{k}, E), \tag{22.2}$$

where $\wp_{\mu\nu k} = e\, d_{\mu\nu k}$ is the transition dipole moment between the subbands, ν, μ, Ω is the sample volume and e is the electron charge. The Coulomb interaction is the origin of the coupling term $V_{kk'}^{\nu\mu}$. Summation over all possible subbands takes into account both resonant, or rotating wave approximation (RWA) and counter-rotating contributions (CRWA). The nonequilibrium population difference between states (or subbands) ν and μ is $N_{\nu\mu k} = N_\nu(k) - N_\mu(k)$. The real and imaginary terms of the selfenergies characterizing the interacting system yield. Introducing auxiliary matrix elements $\chi^0_{\nu\mu}(k, E)$, $\mathscr{F}^{\nu\mu}_{kk'}$,

$$\chi^0_{\nu\mu}(\boldsymbol{k}, E) = \frac{\wp_{\nu\mu k} N_{\nu\mu k}}{E - \varepsilon_{\nu\mu k} + i\Gamma_{\nu\mu k}}, \mathscr{F}^{\nu\mu}_{kk'} = \delta_{k-k'} - \frac{N_{\nu\mu k}}{E - \varepsilon_{\nu\mu k} + i\Gamma_{\nu\mu k}} V^{\nu\mu}_{kk'}, \qquad (22.3)$$

the solution $\chi_{\nu\mu}(k, E)$ is obtained by multiplication by the inverse matrix $\left(\mathscr{F}^{\nu\mu}_{kk'}\right)^{-1}$

$$\chi_{\nu\mu}(\boldsymbol{k}, E) = \sum_{k'} \left(\mathscr{F}^{\nu\mu}_{kk'}\right)^{-1} \chi^0_{\nu\mu}(\boldsymbol{k}', E). \qquad (22.4)$$

22.3 Results and Discussion

Figures 22.1, 22.2 and 22.3 show details of the evolution of both real and imaginary parts of the optical susceptibility and the corresponding α factor spectrum and it's value at peak gain for selected excitation conditions. In all cases, there are more carriers at lower subbands than at the upper bands. Only 3 subbands are considered and labelled $\{e_1\ e_2\ e_3\}$. Details of the notation and the method to obtain the bandstructure of the 7 nm $Ga_{0.98}N_{0.02}As$-$Al_{0.3}Ga_{0.7}As$ quantum well considered here are given in Ref. [13].

Figure 22.3 shows the corresponding evolution of the linewidth enhancement factor. The ensemble of results in the figures above further illustrate the effect predicted in Ref. [8]. The change in occupation of the lower subbands in this range has a slow impact on the α factor structure, but the selective band filling of the lower bands reduces the energy gap between the upper and lower subbands and consequently the position of the peak gain. As a consequence α at peak gain moves up along the steep low energy side of the spectrum leading to a fast increase.

22.4 Conclusion

In summary, this paper further illustrates the evolution of the α factor under lasing without inversion conditions in dilute nitride quantum wells. Regardless of a large chance in the real and imaginary parts of the optical susceptibility, the α factor spectrum remains essentially unaltered. The red shift of the peak gain however makes α at peak gain increase sharply, but still within the order of magnitude of measured values found in the literature for intersubband emitters.

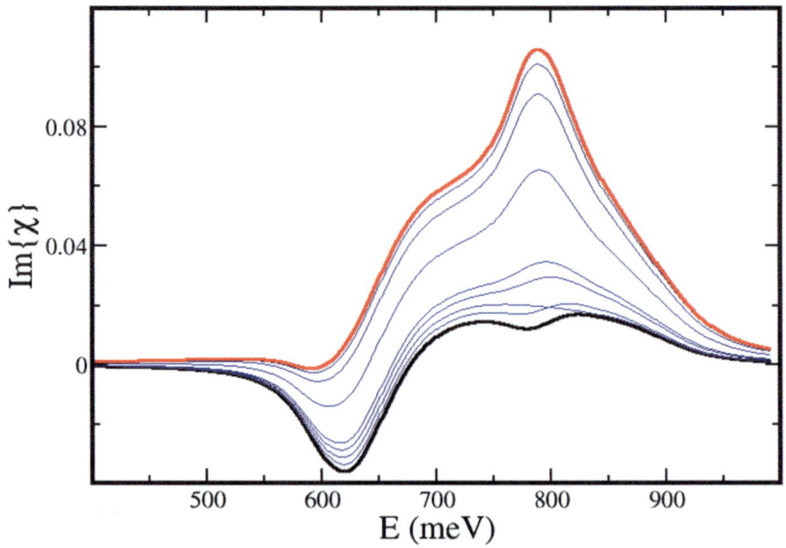

Fig. 22.1 Imaginary part of the optical susceptibility of a 7 nm $Ga_{0.98}N_{0.02}As$-$Al_{0.3}Ga_{0.7}As$ quantum well. In all curves the upper (e_3) subband is occupied with a carrier density $N_{up}=10^{11}\ cm^{-2}$. The lower (e_2) and (e_1) subbands have equal occupation, increasing from the lower (black) to the upper (red curves) by $N_{low} = 1, 1.1, 1.2, 1.3, 1.4, 0.2, 0.25, 0.276$ and $0.28 \times 10^{11}\ cm^{-2}$

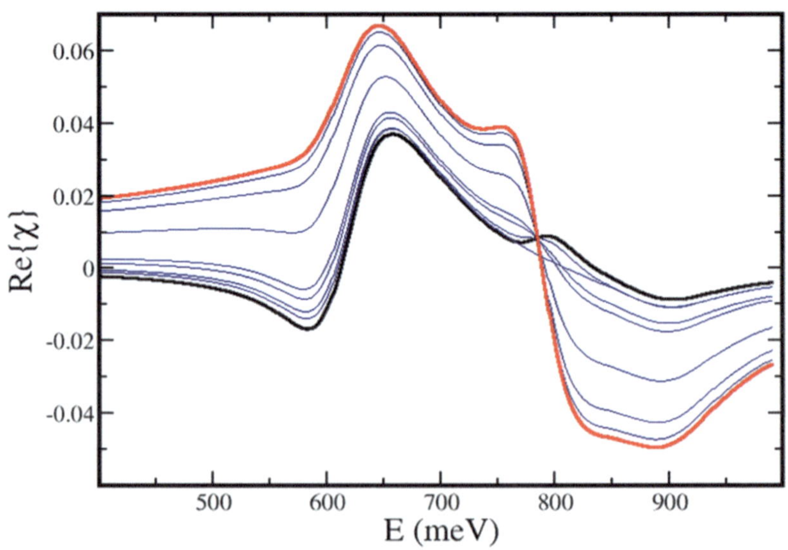

Fig. 22.2 Corresponding real part of the optical susceptibility of the same 7 nm $Ga_{0.98}N_{0.02}As$-$Al_{0.3}Ga_{0.7}As$ quantum well under using the same conventions

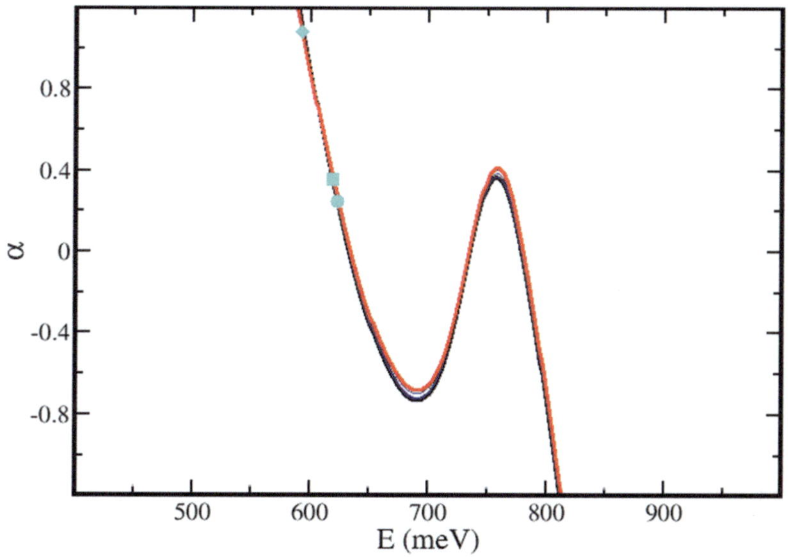

Fig. 22.3 Linewidth enhancement (α factor) for the same sample and excitation conditions of Figs. 22.1 and 22.2. The general structure and numerical value of α factor remains essentially unaltered for all curves. From bottom to top, the circle, square and diamond (cyan) symbols denote the value of α at peak gain, respectively for $N_{low} = 1, 1.4$ and 0.28×10^{11} cm^{-2}

References

1. A. L. Schawlow and C. H. Townes, Infrared and Optical Masers, Phys. Rev. 112 (6), 1940–1949 (1958).
2. C. H. Henry, Theory of linewidth of semiconductor lasers, IEEE J. Quantum Electronics QE-18, pp. 259–264 (1982).
3. M.F. Pereira Jr., S.W. Koch and W.W. Chow, Effects of strain and Coulomb interaction on gain and refractive index in quantum-well lasers J. Opt. Soc. Am. B10, 765–773 (1993).
4. M. Lerttamrab, S.L. Chung, C. Gmachl, D. L Sivco, F. Capasso and A.Y. Cho, Linewidth enhancement factor of a type-I quantum-cascade laser, J. Appl. Phys. 94, 5426–5428 (2003).
5. J. von Staden, T. Gensty, M. Peil, W. Elsaesser, G. Giuliani, and C. Mann, Measurements of the α factor of a distributed-feedback quantum cascade laser by an optical feedback self-mixing technique, Opt. Lett. 31, 2574–2576 (2006).
6. L. Jumpertz, F. Michel, R. Pawlus, W. Elsasser, K. Schires, M. Carras, and F. Grillot, Measurements of the linewidth enhancement factor of mid-infrared quantum cascade lasers by different optical feedback techniques AIP ADVANCES 6, 015212-1–015212-7 (2016).
7. R. P. Green, J.-H. Xu, L. Mahler, A. Tredicucci, F. Beltram, G. Giuliani, H. E. Beere, and D. A. Ritchie, Linewidth enhancement factor of terahertz quantum cascade lasers, Appl. Phys. Lett. 92, 071106-1–071106-3 (2008).
8. M.F. Pereira, The Linewidth Enhancement Factor of Intersubband Lasers: From a Two-Level Limit to Gain without Inversion Conditions, Applied Physics Letters 109, 222102-1–222102-4 (2016).
9. M.F. Pereira, TERA-MIR radiation: materials, generation, detection and applications II, Opt Quant Electron 47, 815–820 (2015).

10. M.F. Pereira and I.A. Faragai, Coupling of THz radiation with intervalence band transitions in microcavities, Optics Express 22, 3439–3446 (2014).
11. M. F. Pereira Jr., Intervalence electric mode terahertz lasing without population inversion, Phys. Rev. B 78, 245305-1–245305-5 (2008).
12. R. Nelander, A. Wacker, M.F. Pereira Jr, D.G. Revin, M.R. Soulby, L.R. Wilson, J.W. Cockburn, A.B. Krysa, J.S. Roberts, and R.J. Airey, J. Appl. Phys. 102, 113104-1–113104-5 (2007).
13. M. F. Pereira and S. Tomić, Intersubband gain without global inversion through dilute nitride band engineering, Appl. Phys. Lett. 98, 061101-1–061101-3 (2011).
14. M.F. Pereira Jr., Microscopic approach for intersubband-based thermophotovoltaic structures in the THz and Mid Infrared, JOSA B28, 2014–2017 (2011).
15. T. Schmielau and M.F. Pereira Jr., Nonequilibrium many body theory for quantum transport in terahertz quantum cascade lasers, Appl. Phys. Lett. 95, 231111-1–231111-3 (2009).
16. T. Schmielau and M.F. Pereira, Microelectronics Journal 40, 869–871 (2009).

MIX
Papier aus verantwortungsvollen Quellen
Paper from responsible sources
FSC® C105338

If you have any concerns about our products,
you can contact us on
ProductSafety@springernature.com

In case Publisher is established outside the EU,
the EU authorized representative is:
Springer Nature Customer Service Center GmbH
Europaplatz 3, 69115 Heidelberg, Germany

Printed by Libri Plureos GmbH
in Hamburg, Germany